Unusual telescopes

UNUSUAL TELESCOPES

Peter L. Manly

CAMBRIDGE
UNIVERSITY PRESS

Published by the Press Syndicate of the University of Cambridge
The Pitt Building, Trumpington Street, Cambridge CB2 1RP
40 West 20th Street, New York, NY 10011–4211, USA
10 Stamford Road, Oakleigh, Melbourne 3166, Australia

First published 1991
First paperback edition 1995

Printed in Great Britain at the University Press, Cambridge

A catalogue record for this book is available from the British Library

Library of Congress cataloguing in publication data available

ISBN 0 521 38200 9 hardback
ISBN 0 521 48393 X paperback

**Transferred to
Digital Reprinting 1999**

**Printed in the
United States of America**

Contents

Illustrations

Preface

This book is about the art and technique of telescope design, although to a certain extent it comes under the heading of "astronomical trivia". While complete descriptions of the designs are not possible in a single volume, more information can be obtained from the reference lists for each telescope. The book isn't about just slightly unusual telescopes as shown in Fig. 1 – it's about some blatantly odd instruments such as that shown in Fig. 2. These are mystical machines. They are often at the leading edge of technology. They are not as unusual, however, as the student of a far Eastern philosophy who told me that she focused the whole life force of the Universe to peer into other and much more distant planes of existence. In this book, we're not going to travel that far afield.

In order for astronomy to be successful, there must be observers, theoreticians, instrumentation engineers, writers, technical managers and even janitors who maintain the astronomical industry (and if you don't think janitors are important, just ask a professional observer how much work gets done when the drains back up on Kitt Peak).

This book is dedicated to the astronomers who have an interest in the machines, devices and techniques of observation. It is dedicated to the amateurs and professionals who have designed telescopes. They may have created new telescopes to further the state of the art in high technology scientific endeavor or they may have designed a telescope simply because they were too poor to buy a commercially made instrument. They may have brought forth a new type of telescope to make a specific observation which was not possible on any existing telescope. Perhaps nobody had ever thought of making that kind of observation before. Most important, the new telescope often resulted in an observation which led to an understanding of some phenomenon which clarified our perception of the wonderful Universe in which we live. Such an observation may actually clarify a little, but it often opens up even more questions which still must be answered. Such is the nature of science.

The telescope is the prime tool of the observational astronomer. Yes, I know that the computer is also a big tool. Let somebody else write about astronomical computers. The theoreticians may want to lynch me but I feel the telescope is, indeed, the prime tool of astronomy, especially for amateur observers. Its history

Figure 1. A slightly unusual telescope. While the telescope is functional, it is slightly out of proportion with its base and the truss structure is too long. The balance also appears to be a problem.[1] Photograph courtesy of Steve Coe, Saguaro Astronomy Club.

since Hans Lippershey[2] or some other early seventeenth-century designer discovered the principle, has been colorful. In nearly four centuries, thousands have been built for many purposes. The first crude instruments have been reduced to exacting practice by hundreds of engineers, observers and tinkerers. As tools of science and technology, telescopes often incorporate state of the art advancements in optics, mechanics and control systems.

It's hard to classify unusual telescopes when each of the illustrations presented here is already categorized "Miscellaneous". There are some instruments about which you'll ask "That's a telescope?". Indeed, they are all telescopes, although some are telescopes only by stretching the definition. Most of the telescopes were made to serve the desirable purpose of observing the skies. Some were obviously made to allow the owner to possess an impressive looking machine. Some show the careful craftsmanship indicative of a lover of well-made scientific instruments. These

[1] The evolution of this telescope is described in Telescope Making, No. 26, Summer, 1985, p. 10. The telescope was used in this form for only two nights before it was rebuilt.

[2] The History of the Telescope, Henry C. King, Dover Publications, 1955. There has been considerable discussion concerning the discovery of the telescope. A short summary of various claims can be found in an article by Brian Slade, FRAS, in Telescope Making, No. 30, Summer, 1987, p. 44. The subject is also discussed in The Telescope, Louis Bell, Dover Publications, 1981 edition, p. 3.

Figure 2. The Paris Observatory Coudé Refractor. Illustration courtesy of Paris Observatory. Note: a complete description of this telescope is covered in Chapter 7.

devices, with their hand-tooled brass knobs, polished hardwoods and functional design are a work of art in the eye of a mechanical engineer. Others appear as if they were thrown together by a demented scrap dealer (indeed, one telescope often seen at the Riverside Telescope Makers Conference was used for a prop in a TV show, masquerading as a laser cannon operated by a mad scientist).

Some telescope designers simply march to the beat of a different drummer. They want a machine which looks different just to be different. One telescope designer admitted that he only wanted a telescope that didn't look like any other telescope in the astronomy club — and especially not like one of those orange store-bought things.[3]

This is very much a story of honest engineering attempts to push forward the state of the art in astronomical instrumentation. Most of these inventions actually worked — after a fashion. Some didn't but when they didn't, at least we learned something (we learned how not to make a telescope). It is instructive to the telescope designer to see some of the more notable failures. I will try to point out design flaws gently, for I have produced my own flawed designs in the past and I understand the receipt of criticism. Many of the ideas in telescope design presented here looked, at

---

[3] I have nothing against those orange store-bought things. In fact, I own two of them and use them regularly. Of course, I've tinkered with the design a little.

first glance, like quantum leaps in the state of the art. Alas, some new engineering concepts were like a Greek tragedy in that they included a fatal flaw. We'll see at least one design, multiple mirrors, in which the fatal flaw was surmounted decades later by new mirror-support technology and the concept worked beautifully.

There is no logical starting point in a subject where everything is miscellaneous so we'll just dive right into the middle. We'll consider telescopes that work in optical, infrared (IR), radio and every other portion of the electromagnetic spectrum.

In rough order, we'll see optics, telescope mounts, limits (both large and small) and, of course, miscellaneous telescopes. While researching this book I was surprised to find what I thought were some radical designs — only to discover later that the same design had been invented 20, 40 or 100 years earlier. Indeed, some strange concepts appear to have a reincarnation period of about 20 years, judging by publications like *Sky & Telescope*, the *Journal of the British Astronomical Association* and *Astronomy*. There are some names and organizations which surface often such as Léon Foucault, Russell Porter, Richard Buchroeder, Donald Dilworth, John Dobson, Oscar Knab, John Wall, Joe Perry, the Riverside Telescope Makers Conference, Stellafane, the David Levy telescope collection and just about any telescope associated with the Paris Observatory.

A note to the reader; if you see your telescope listed here and are offended that I have called it unique, please accept my apology. It must have been at least slightly unusual to come to my attention. On the other hand, if you don't see your telescope here and are offended because you think it's unusual then please write me. I can be reached at;

1533 West 7th Street, Tempe, AZ 85281–3211, USA

You can try to reach me on the phone but most clear evenings I'll be out by the telescope — and most evenings in Arizona are clear. You will also have the problem of getting through the busy signals, for I have teenage children who believe telephone privileges are part of the Bill Of Rights. I should be reachable by phone in around 1996 when the youngest will be away in school.

A note to my US friends; please do not write me to say that I spell words funny. My publisher allowed me to write the book using North American words. While editors are usually very understanding about most things, they may add text during final proof reading using words such as color (colour), program (programme), etc. In essence, I have been dragged kicking and screaming from the outback of the Great American Desert into civilized British publishing standards.

Acknowledgements

I would like to thank David Levy, who convinced me to tackle a book by myself. There are almost a hundred people who contributed photographs, provided descriptions of telescopes and gave me leads to even more unusual telescopes. Most of them are mentioned under each telescope but they deserve acknowledgement here too. There were, however, several who took an avid interest in the book, dug up references and searched out old photographs. They include, in no particular order, Dr Clyde Tombaugh, Richard Buchroeder, John Wall, Berton C. Willard, John F. Martin V, Dan Brocious and Oscar Knab. The members of the Saguaro Astronomy Club, the East Valley Astronomy Club, the Phoenix Astronomical Society and the Writer's Refinery were instrumental in critiquing the subject matter. I would also like to thank Rick and Patti Cook and Leroy and Frances Paller for providing computer and printing assistance at critical times.

Then there is my family, who have refrained from bothering me while providing a never ending supply of coffee and munchies.

The manuscript for this book was written on Macintosh™ 512+ and Macintosh™ SE computers operating with a Crate Technology® 60 Megabyte hard drive. Both MacWrite™ and Microsoft® Word were used to create text while MacDraw™, MacDraft™ and MacPaint™ software were used to create drawings. Some illustrations from old or damaged photos were scanned into the computer using a ThunderScan® picture digitizer and then cleaned up with GIFConverter© image processing software. I would never have attempted writing a book without all of that hardware and software.

1

Optics

In this chapter, several different mirror and lens materials will be considered. No matter what the material, however, it must be shaped into a smooth curve by grinding, polishing or some other fabrication technique. Astronomers say that it is a mark of intelligence to be able to grind a good mirror. It has also been quipped that a mark of stupidity is to grind a second one. Mirror grinders and other glass pushers become attached to their work, having spent many loving hours grinding a precise optical figure into the surfaces, testing, putting a final stroke or two on the glass, testing, making just one more correction, testing, getting it just right, testing. Mirror grinding (figuring) is an occupation for the terminally finicky. It is no wonder that cutting remarks can fly at a meeting of opticians. For instance, I am reminded of something overheard at the Riverside Telescope Makers Conference; "What'd ja polish that thing with, boy, peanut butter and a brick?"

Rock, clay and ceramic mirrors

The primary component in telescopes is always either the mirror or the objective lens. Traditionally, these have been made of glass. Remember, however, that the original mirrors by Newton, Herschel and Lord Rosse were of speculum metal. A chief attribute of optics, no matter what the material, is that the optics be rigid and stable. Floppy disks may work in computers but floppy optics have seldom been popular in astronomy. Mirrors have been fabricated of glass, ceramic, even granite and obsidian.[4]

Obsidian and granite are clever mirror materials. Formed as molten ingots much like glass, these materials were cooled or annealed slowly over centuries and thus their internal grain structures have very few stresses. Several people have made mirrors from these materials but they report that obsidian and granite are much harder than glass and require considerably more work to polish correctly. Obsidian has a higher thermal conductivity than glass and thus the mirror tends to come to temperature more rapidly at the start of an observing run.[5] Chris Pratt of the San Jose

[4] Obsidian mirrors are described in *Sky & Telescope*, August, 1981, p. 122 and November, 1979, p. 410.
[5] *Gems & Minerals*, June, 1966, p. 25.

Figure 1.1. 20.32 cm (8 in) f/8.2 obsidian primary mirror before silvering. Note the grain of the obsidian. The mirror was made by Jackson T. Carle. Photo courtesy of Chris Pratt.

Astronomical Association reports very good results with the mirror shown in Fig. 1.1. The mirror was finished with an aluminum reflecting surface using conventional vacuum coating techniques.

Donald Dilworth tried a slightly different approach to making a mirror. He formed clay into the correct approximate shape and then fired the clay with a glaze to produce a rigid ceramic mirror which could then be polished to the correct figure.[6] While the technique has not proved fruitful, many modern mirrors are being cast from temperature insensitive ceramics such as Zerodur®.[7]

The Astro Met Corporation has produced open cell ceramic foam mirror materials. The foam is much lighter than conventional glass and since air can flow through it, the mirror will stabilize in temperature much more rapidly. For the current technology, a thin aluminum front surface is bonded (glued) to the ceramic foam core and then the metal is polished. In the future, nonporous ceramic mirror materials for the reflecting surface may be formed with a foam ceramic backing. This would result in a very stable, lightweight mirror which responds quickly to temperature changes. Currently, the ceramic foam mirror is still in the research and development stage.

[6] Dilworth's ceramic mirrors are discussed in *Sky & Telescope*, October, 1975, p. 259.
[7] A ceramic mirror by Patric Canan was shown at the 1983 Riverside Telescope Makers Conference. The 24 cm (9.5 in) f/3.8 mirror is described in *Telescope Making*, No. 20, Summer/Fall, 1983, p. 18.

Glass foam has also been used as a lightweight backing material in some mirrors. In most applications, a closed cell foam is used which acts like very hard Styrofoam®. In a typical application, a 1 mm thick flat sheet of silvered glass about a meter in diameter is bonded (glued) to the glass foam backing which has been pre-formed into the required concave parabola. The thin glass sheet is kept in contact with the curved glass foam backing by vacuum suction until the epoxy bond has cured. In order to protect the aluminum reflecting coat from the weather, it is usually placed on the back side of the mirror, much like a Mangin mirror. High precision optical surfaces have not been attainable with this method and mirrors made using this technique have usually been employed as solar collectors. There is, however, a class of low angular resolution, large aperture astronomical telescopes used to detect cosmic-rays and gamma-rays which use these mirrors extensively. Often hundreds of these mirrors are arrayed into a collector with an effective aperture of about 10 m (33 ft). These telescopes are discussed in the section on noncoherent multiple mirror light buckets in chapter 8.

Plastic optics

Optical fabrication experiments have been made with plastic and similar materials such as Perspex™.[8] English astronomers John Wall and E. Dunlop have cautioned, however, that the plastic materials should be annealed before grinding and that the long-term stability of plastic optics is not yet established. Wall ground and polished a 0.91 m (36 in) f/12 aperture lens for a Schupmann (medial) refractor. After final polishing, however, the surface appeared rough under a Foucault test even though it appeared well polished to the eye. The project was abandoned but the lens still exists. Perspex™ and other plastics also have the disadvantage that they turn yellow after a couple of decades, discouraging their use for lenses. The original reason for using plastic — a lesser expense than glass — has now evaporated with rising prices for large plastic sheets. Wall also made a 15 cm (6 in) aperture f/3 mirror and found it easy to silver-coat the plastic.

In recent years the advent of cheap refractors with molded plastic lenses has given a bad name to the use of this material. Plastic lenses and mirrors larger than 10 cm (4 in) in aperture have not been made in quantity. There is a gray area as to where plastics end and composite materials, discussed below, begin.

Metal and composite material mirrors

Speculum metal, an alloy of copper, tin, occasionally zinc and traces of arsenic, was the first choice of mirror makers. Long before glass-silvering technology was developed the white, hard metal was used to make reflecting surfaces. The exact

---

[8] *Journal of the British Astronomical Association*, Dec 1977 Vol.88, No. 1 p. 28 and Aug 1978 Vol.88, No. 5 p. 517.

mixture ratios of the ingredients were often kept secret by mirror makers, as were the identity of other seasonings in the recipe. William Parsons, the third Earl of Rosse, tried several exotic techniques in casting large metal disks over 0.5 m (19 in) across. He experimented in the late 1800s with soldering speculum metal plates on a solid brass backing fixture and then grinding the whole assembly to the proper curve.[9] Because of diffraction problems associated with the joints of the plates, he dropped the idea and developed methods for casting large specula, up to about 2 m (6 ft) in diameter.

In general, metal mirrors have problems maintaining the required dimensional stability over a normal range of temperatures. On the other hand, they can be made much thinner and thus more lightweight than glass. This is because speculum metal is stiffer and stronger than glass, which is really just a very viscous liquid.[10] There are some applications in which metal mirrors are required, such as in spacecraft where weight is a premium.[11] Beryllium is often used instead of aluminum because beryllium is stiffer and has better dimensional stability. It is, however, a hazardous metal with which to work since beryllium dust and metal shavings are highly poisonous.

Until recently, fabrication of glass mirrors over 5 m (16 ft) in diameter was nearly impossible. Thus, for those applications involving lesser requirements for optical surface smoothness, as in infrared and radio applications, metal mirrors were considered because of the low cost involved in making large metal surfaces.

As an example, the laser-ranging telescope of the Tokyo Observatory uses a 3.8 m (150 in) metal dish as shown in Fig. 1.2. Note that the secondary support struts pierce the primary mirror, a feature not often seen on glass telescope mirrors.[12] This telescope, which has the same aperture as the Mayall Telescope, largest telescope at Kitt Peak, was built for a fraction of the cost. While the metal mirror telescope has significantly less resolution than a glass mirror, the requirements of laser ranging are simply that the mirror act as a large aperture, not as a precision imaging system. Telescopes designed for maximum aperture are often referred to as light-buckets.

With the advent of exotic materials designed for the space and defense industries, some of these were bound to find their way into telescopes. Daniel Vukobratovich of the University of Arizona Optical Sciences Center has designed a 30.48 cm (12 in) aperture, 4.5 kg (10 lb) telescope made largely of metal matrix composites; as shown in Fig. 1.3. The structural members are made of SXA, a new class of dimensionally stable composites made of aluminum and fine silicon carbide particles. The mirror is

---

[9] The History of the Telescope, Henry C. King, Dover Publications, 1955, p. 207.

[10] The fact that glass is a viscous liquid can be demonstrated by measuring the thickness of a window pane in an old house. The bottom will be thicker than the top because the glass, over a period of decades, is flowing to the bottom of the window. It has been theorized that glass mirror or lens telescopes which are always stored on a particular side should develop an astigmatism after a century or so. The effect, however, has never been observed.

[11] A typical beryllium mirror for use in an infrared spacecraft telescope is described in Sky & Telescope, November, 1976, p. 340.

[12] A description of the Tokyo laser-ranging telescope is in Sky & Telescope, November, 1975, p. 280.

Figure 1.2. Tokyo Observatory 3.8 m (150 in) metal mirror telescope used for lunar ranging with a laser. Photo courtesy of Yoshihide Kozai, National Astronomical Observatory, Mitaka, Tokyo.

made of a sandwich of metal foam with solid metal plates for the mirror and back plate. The design is interesting in that the primary mirror has a curve formed into it on both the front and the back sides. The reason for this is so that the slump of the mirror due to self weight and the change in curvature due to thermal stresses are minimized.[13]

In most conventional mirrors, the light strikes the surface almost at right angles. Photons of higher energy, however, will not focus properly with this type of mirror. In fact, they will probably penetrate the mirror surface and be absorbed rather than reflected. Thus, in the extreme ultraviolet and soft X-ray regions photons are difficult to focus. These higher energy photons will reflect if they strike the mirror at a shallow or grazing angle, say less than about 5°. A reflecting mirror can then be built in which the photons just barely graze the surface.[14] Since the cross-section of the collecting aperture of the telescope tends to be a narrow annulus, the effective collecting area is rather small. This can be increased by nesting concentric mirrors inside one another as shown in Fig. 1.4.

Each of the annular segments is actually a small section of a deep parabolic mirror with all of the mirrors having a common focus. The assembly could be considered a multiple mirror telescope design taken to the extreme. The completed telescope thus

[13] A description of the lightweight telescope is in *Optical Engineering*, the journal of the Society of Photo-Optical Instrumentation Engineers, February, 1988, p. 97.
[14] Grazing incidence optics are discussed in *Sky & Telescope*, January, 1969, p. 14 and May, 1969, p. 300.

Figure 1.3. Lightweight telescope of 30.48 cm (12 in) aperture weighing only 4.5 kg (10 lb). Photo courtesy of University of Arizona.

resembles a cookie cutter for making extremely thin donuts. The requirement that as many mirrors as possible be nested implies that the mirrors should be as thin as possible. Typically, metal optics have been used extensively in this application since glass optics would need to be much thicker in order to retain the desired stiffness.

It is interesting to note that the mirrors are made by machining the metal on a special highly accurate diamond turning lathe. Mirrors such as this have also been made with a machined surface of lithium fluoride. X-rays diffract when encountering this crystal by means of Bragg diffraction, thus forming a diffracting, rather than a refracting telescope. Since the Earth's atmosphere absorbs most of the extreme ultraviolet and soft X-rays, these telescopes have typically been used on either balloon-borne payloads or on spacecraft.

Glass/metal mirrors

In an effort to decrease the costs of glass on larger telescope mirrors, several attempts have been made to use either thin mirrors or mirrors made up of several smaller

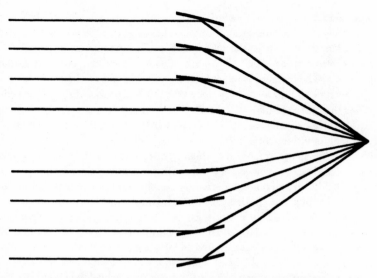

Figure 1.4. X-ray telescope composed of several nested grazing incidence mirrors.

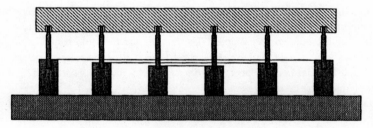

Figure 1.5. Hairbrush mirror mount.

pieces. Thin mirrors have the problem that they are not dimensionally stable as the telescope moves around the sky unless elaborate mechanisms such as multi-point flotation cells and complex edge supports are used to keep them from deforming under their own weight. In telescopes over a meter in aperture, these mechanisms can drive the cost of the telescope out of bounds.

Eric Mobsby of Dorset, England, has taken a different approach. He has made very thin mirrors but has attached them with epoxy to a cast aluminum back plate which has many rods supporting the mirror, as shown in Fig. 1.5. Eric calls this the hairbrush mirror mount and the idea is to let the aluminum casting expand at its own rate which is slightly different from the glass. The short vertical rods will then bend slightly to make up the difference. There are three concentric rings of rods and each base ring is cast into a backing structure which allows air to circulate freely through the assembly.[15] This technique should not be confused with active mirror supports in which the length of each of the metal rods might be adjusted independently to control the mirror surface. Such mirrors are covered in the section on flexible mirrors in this chapter. The metal rods in Mobsby's mirror support are bonded with epoxy into shallow holes drilled in the back side of the thin mirror. Grinding and polishing is done after the mirror blank is attached to the casting. His mirrors, which are up to 30.5 cm (12 in) in diameter and about 1 cm (0.4 in) thick have been subjected to some rather brutal environmental tests (how many people deliberately leave their telescope mirrors in the freezer for a while?). The results to date have been good but replacing a complex mirror cell and support mechanism with a complex custom casting probably isn't advantageous unless volume production methods are used.

Eric Mobsby had earlier tried making lightweight mirrors from glass top and bottom plates, separated by either glass or metal spacers. The problem usually encountered was that the differential expansion between the glass plates and the metal spacers was excessive. Corrugated metal spacers were tried on the assumption that the folded metal would bend sufficiently to accommodate the differential expansion.[16] This type of construction has proven successful if the plates and spacers are made of the same material and fused together, as has been done for years with fused silica and more recently fused Pyrex®.[17]

G.W. Ritchey made several mirrors up to 1.52 m (60 in) in diameter by glueing glass support ribs to the back of a single thin primary mirror blank which had been rough-ground to the correct shape.[18] The mirrors did not hold together for more than a few years before the glue deteriorated. Recent attempts to epoxy glass plates side by side in order to make larger mirrors have met with mixed success, largely due to problems at the epoxy joints.[19]

In a similar effort, Dr Sherman W. Schultz of Macalester College, Minnesota, is assembling a 0.86 m (34 in) f/3 mirror as shown in Fig. 1.6. The original objective in his explorations into mirror fabrication technology was to decrease the quantity of

[15] The hairbrush mirror mount is described in Sky & Telescope, December, 1978, p. 569.

[16] A description of Mobsby's experiments with epoxied mirrors is in Sky & Telescope, November, 1964, p. 305.

[17] Some of the original development of the techniques for fusing Pyrex ® blanks is described in an article by John M. Hill and Roger Angel in Telescope Making, No. 14, Winter, 1981/1982, p. 24. The Hextek Corporation of Tucson, Arizona and Star Instruments, Inc. of Flagstaff, Arizona have pioneered in the technology of producing fused Pyrex ® mirror blanks. For more information, see Sky & Telescope, July, 1984, p. 71 , January, 1986, p. 100 and December, 1988, p. 698.

[18] The Development of Astrophotography and the Great Telescopes of the Future, G. W. Ritchey, Astronomical Society of France, 1929.

[19] As with Rosse's attempts to make large speculum metal mirrors from several smaller ones, the joints are not completely smooth between adjoining plates. These joints cause diffraction and thus decrease the effectiveness of the mirror in terms of resolution. See Sky & Telescope, December, 1984, p. 558.

Figure 1.6. Primary mirror built up by epoxying glass plates together. Photo courtesy of Sherman Schultz.

glass which had to be removed by grinding. This is especially important with lower f number systems where telescope makers joke that they are making salad bowls.[20] Sherman's approach was to assemble the mirror from several 30.5 cm (1 ft) square plates side by side. The plates were first rough-ground to the desired spherical shape and then assembled face down on a form with the correct convex curvature. The separate plates were then bonded together with a special epoxy for glass. On the back side of each joint there is a fiberglass strap epoxied over each joint to add strength.

The assembled mirror was then ground with a sub-diameter tool. Initial results, while promising, pointed to the need for more support on the back side since the

[20] A novel approach to pouring mirror blanks which do not require heavy grinding to rough out the basic curve has been taken by Roger Angel of the University of Arizona. He makes "spin-cast mirrors" in which the melting and annealing oven rotates like a merry-go-round, allowing the glass to solidify with a parabolic surface, much like the rotating mercury mirrors described in the section on liquid mirrors. Two meter mirrors have been made with this technique and mirrors of up to ten meters are planned.

entire assembly flexed during grinding. More rear supports have been added to the mirror and final grinding is proceeding at the time of this writing.[21]

Flexible mirrors
Liquid mirrors

I mentioned earlier that telescope optics should be rigid. With the advent of flexible and liquid optics (yes, I mean liquid optics) perhaps a more accurate phrase would be that telescope optics should be stable.

Both lenses and mirrors have been made of liquids. In the late 1700s it was well understood that in order to correct lenses for color aberrations, at least two lenses of two different indices of refraction were required. The technology was available to produce relatively large blanks of crown glass but flint glass was scarce in larger sizes.[22] Euler experimented with liquid lenses during this period. Typically, the liquid was used as a refractive medium only and was held in place with meniscus zero power lenses. Robert Blair, a Scottish Naval surgeon, experimented with water and metal salts as a refractive medium. Between 1827 and 1832 Peter Barlow constructed several liquid-filled objectives with apertures ranging from 15.2 cm to 20.3 cm (6 in to 8 in). Tests comparing liquid lenses with conventional glass optics showed the two technologies were competitive.[23] There was always the fear, however, that the lenses would leak or change color with time. In addition, there was concern that some of the acids used as refracting media would etch the glass which contained them. Many of the telescopes, however, were used for years without serious problems. With the advent of more modern glass materials in the early 1800s the development of liquid lenses was abandoned.

A flat pan of mercury has been used for years in zenith telescopes worldwide. A typical example from the US Naval Observatory is shown in Fig. 1.7. The zenith tube is used to observe the meridian passage of stars for timing and geodetic location purposes.[24] It is a requirement that the telescope be pointed absolutely vertical. The pan of mercury, acting as a mirror, is guaranteed to be flat (to within the radius of the curvature of the Earth) and perpendicular to the vertical direction. No adjustment is required. The objective lens in the top of the tube and a downward looking photographic plate holder located just under the lens rotate about the axis of the tube. This rotation allows exposures to be made with one edge of the lens toward the East and then, after rotation, toward the West. Thus, instrument errors are calibrated out of the system in this manner.

[21] A photograph of the partially completed mirror is in *Sky & Telescope*, December, 1984, p. 559. A description of the mirror design effort is in *Telescope Making*, No. 24, Fall, 1984, p. 41.

[22] *The History of the Telescope*, Henry C. King, Dover Publications, 1955, p. 155.

[23] *The History of the Telescope*, Henry C. King, Dover Publications, 1955, p. 190.

[24] The application of a photographic zenith tube telescope is described in *Astronomy*, Robert H. Baker, Van Nostrand Company, Inc, 1964, p. 74.

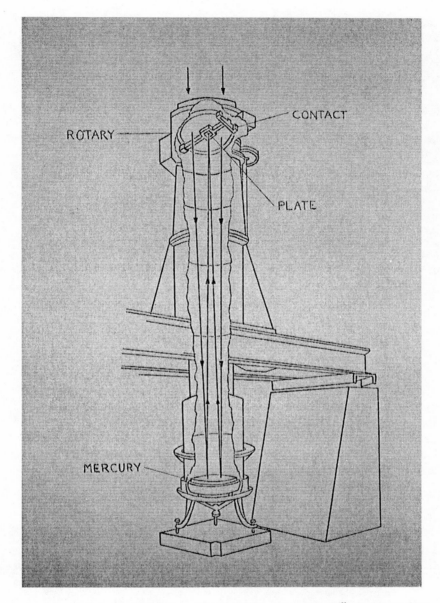

Figure 1.7. Photographic Zenith Tube telescope with mercury mirror. Illustration courtesy of US Naval Observatory.

Figure 1.8. Liquid mirror formed by rotating a pool of mercury. Photo courtesy of
Ermanno F. Borra, Université Laval.

As early as 1908 physicist R.W. Wood made a 51 cm (20 in) focusing mirror using
a rotating pool of mercury in a shallow pan.[25] As the pan rotated, centripetal
acceleration forced the mercury toward the edge to pile up. The curved surface
formed a parabola which just happens to be the desired surface of a telescope mirror.
Technical problems encountered were that the pan had to rotate without vibration
which would cause ripples on the surface and the speed of rotation had to be a
constant, lest the focal length of the system change. After 1909 Wood discontinued
tests on rotating mirrors.

More recently, Ermanno F. Borra built a 1.6 m (63 in) mirror at Laval University,
Quebec, as shown in Fig. 1.8.[26] He employed modern air bearings in order to
eliminate the problems of vibration. Naturally, such a telescope can point only

[25] See the *Publications of the Astronomical Society of the Pacific*, Vol. 99, No. 621, November, 1987, p. 1229.
See also *Astrophysical Journal*, March, 1909 and *The History of the Telescope*, Henry C. King, Dover
Publications, 1955, p. 401.

[26] A description of rotating mercury mirrors is in *Sky & Telescope*, September, 1984, p. 267. A discussion
of applications and testing of the mirrors is in the *Publications of the Astronomical Society of the Pacific*,
Vol. 97, No. 591, May, 1985 p. 454. A description of the telescope with a one meter mirror installed is
in the *Astrophysical Journal*, No. 279, October 15, 1985, p. 846. A proposed six meter rotating mercury
telescope is discussed in the *Publications of the Astronomical Society of the Pacific*, Vol. 99, No. 621,
November, 1987, p. 1229.

toward the zenith but there are several tasks such as supernova searches of selected strips of the sky and long-term spectroscopic inventories which can be performed with a zenith telescope. A 1.22 m (48 in) rotating mercury mirror has been installed in a telescope and images of 0.25 arc second accuracy were obtained over integrations of 5 seconds, which is the period of one full rotation of the mirror. A 1.65 m (65 in) mirror is currently being shop-tested. Since the focal length of the mirror is merely a function of the speed of rotation of the pan, the possibility exists that telescopes of variable focal length are possible. If a long focal length is desired, slow the rotations. If wide fields are needed for the next observation, crank up the motor and move the camera inward. The 1.65 m mirror has been tested with a focal length as short as 1.5 m (59 in) or f/0.91. Mirror cleaning is much easier for this type of telescope. The mercury is drained from the pan and filtered. There are, however, safety and toxicity problems associated with mercury.

Recently a more exotic nonrotating liquid mirror has been proposed by Borra.[27] It consists of a dish of mercury forming a flat surface. Over the dish of mercury is poured a clear liquid with a radial gradient index of refraction. This means that the index of refraction of the transparent liquid is not the same at the center as at the edge of the mirror. This is accomplished by dissolving different amounts of salts in the transparent liquid which is usually water. The water will tend to mix and the denser portions will sink to the bottom. In order to prevent this, a slow but steady flow of liquid must be pumped into the dish at various radii while mixed fluids are drained from the bottom of the dish. All of this must be done without causing a ripple on the surface of the liquid.

Starlight enters the clear liquid and is bent according to the local index of refraction. It reflects off the mercury surface at the bottom of the pool and is bent again as it goes up through the clear liquid. Since the rays traverse the refracting medium twice, the system acts much like a Mangin mirror. Due to the differing indices of refraction across the pool of clear liquid, light striking the edge of the mirror is bent at a slight angle with respect to light striking the center of the mirror. With proper control of the densities and indices of refraction of the transparent liquid, these rays can be brought to a common focus and thus form an image. Theoretically, mirrors tens of meters in diameter could be fabricated using this technology. One of the primary problems in this approach is that the liquid must be kept flowing in order to maintain its radial gradient of the index of refraction. A pump, flow meters, nozzles and drains must be constructed which will keep the liquid in motion without causing a ripple on the surface of the liquid. Ripples might be eliminated with a glass cover, but remember that the original intent of the liquid mirror was to avoid fabricating a large piece of glass.

Common water has also been used to make focusing optics. Several years ago at the Riverside Telescope Makers Conference a device was shown which consisted of

[27] *Publications of the Astronomical Society of the Pacific*, Vol. 99, No. 622, December, 1987, p. 1371 and *Publications of the Astronomical Society of the Pacific*, Vol. 100, No. 630, August, 1988, p. 1015.

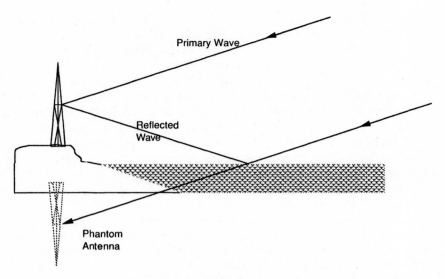

Figure 1.9. Lloyd's mirror.

a hoop with an aluminized mylar sheet attached to the bottom. The mylar was placed horizontally and water was poured into the hoop. As the weight of the water caused the mylar to stretch into a curved surface, a focusing mirror was formed, allowing the light to pass through the water, reflect off the curved mylar and come to a focus above the level of the water. The device was more useful as a solar energy collector but it did work — after a fashion.

A flat sheet of water can be used to make a mirror for a radio telescope as shown in Fig. 1.9. In a simple form, many radio telescopes are a pair of antennas separated by a known distance. Radio signals from both antennas are summed and, depending on the radio wavelength and the antenna separation, the signal from one antenna will interfere with (by either summing or cancelling) the other signal. As the source of the signal moves with the sky, the geometry of the antennas with respect to the source changes and the signals will alternately sum and cancel each other. This is the classical "fringe" pattern used to determine radio source sizes and point source separations. In order to eliminate the cost of one antenna, a single antenna may be placed on a cliff above the sea. On calm days, when the water roughness is less than the radio wavelength of interest, the sea acts as a mirror, creating a phantom antenna below the real antenna on the cliff. The signal output of the real antenna acts as if there were two antennas with summed outputs. This arrangement, known as Lloyd's mirror[28], was used in the early days of radio astronomy. The technique has been supplanted by two-dimensional arrays of fully steerable parabolic antennas.

[28] *The History of the Telescope*, Henry C. King, Dover Publications, 1955, p. 438. A photograph of the installation at Dover Heights, near Sydney, is in *Sky & Telescope*, January, 1953, p. 60.

The use of a calm lake as a mirror has been suggested to photograph bright meteors near the horizon.[29] If a camera were placed at the lake shore with about half of the field of view showing the lake, then two images of a meteor would appear. One would be the direct image of the meteor in the sky and the other would be the reflected image which is the equivalent of viewing the meteor from a vantage point below the water. The location of the two meteor path images would be slightly different with respect to the background stars. This might allow triangulation of the meteor position with a single frame of film. Since the displacement would be very slight and the lake surface must be extremely smooth, a better method for obtaining triangulation is the standard technique of using two separate cameras separated by a few kilometers.

Membrane mirrors

Stretched aluminized mylar was used in a telescope by Maurice V. Gavin of Surrey, England. He partially evacuated the volume behind the mylar with a common household vacuum cleaner and thus produced a curved mirror.[30] The 53 cm (21 in) aperture f/1.9 telescope was built from scrap parts. The focal length, of course, depended on the amount of vacuum applied. Mr Gavin has tried mirrors with f numbers as low as 0.25 but he finds that spherical aberration below f/1 is excessive. The mirror, using commercial grade aluminized mylar, does not have a good enough surface for visual observing but the 5 mm images are small enough for photometric detectors, where all that is required is that the telescope be a light-bucket. The telescope has been used in infrared and solar observations as a flux collector.

Dr Peter Waddell of the University of Strathclyde, Scotland, used a similar procedure to make image quality mirrors up to 96 cm (26 in) in aperture as shown in Fig. 1.10. He found that a concerted effort must be made to keep the rim of the mirror support smooth and free of small nicks or protrusions, lest these irregularities deform the mirror surface as shown in Fig. 1.11. This implies that while the mirror material is less expensive than ground and polished glass, the mirror cell is much more complex than a conventional mirror support. In addition to having a sealed rear surface, the cell's front edge must be carefully machined and polished where it contacts the mylar. It must also have a mechanism which clamps the mylar firmly without stretching or deforming the material. Care must be taken when selecting the mylar material. Sections that have been stressed or pulled will stretch unevenly when the vacuum is applied. These edge distortions could possibly be masked off by placing an aperture in front of the mirror which blocks the outer 10% of the mirror surface.

Since the focal length of the mirror is a function of the differential pressure on either side of the mirror membrane, the vacuum pump must be well regulated. The principle of maintaining a pressure differential will also work for a telescope

---

[29] *Sky & Telescope*, September, 1989, p. 328.
[30] The telescope is described in *Sky & Telescope*, May, 1979, p. 489.

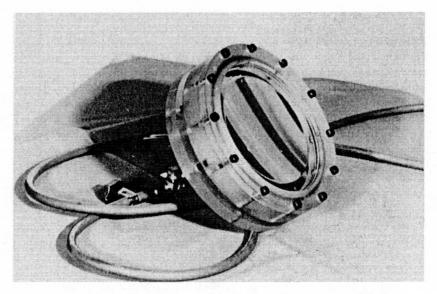

Figure 1.10. Vacuum-formed aluminized mylar mirror. Photo courtesy of Dr Peter Waddell.

pressurized from the inside with the membrane open to the air at the rear. While such an inflatable telescope also requires a clear membrane at the front to keep the pressure in, the design would be useable in the vacuum of space.

Even though the shape of a mylar membrane with a vacuum pulled on it approximates a parabola, variations in stretching properties from one place on the surface to another will cause small imperfections. A clever solution to this problem was developed by Jim Faller of Colorado. He mounted high voltage electrode rings near the back surface of the mirror and parallel to it. By varying the voltages on the electrodes, small electrostatic attractive and repulsive forces were generated between the electrodes and the mirror, thus deforming the mirror slightly and tweaking the final figure of the mirror.

Flexible glass mirrors

Guillermo Cock of Medellin, Colombia employed a similar vacuum concept in making a very long focus 26 cm (10.2 in) aperture Cassegrain as shown in Fig. 1.12.[31] He used a common thin flat window plate about 5 mm (0.2 in) thick, deformed by vacuum to make the mirror. Only mirrors with fairly high f numbers such as his f/15 telescope can be made this way. This very thin mirror weighs only a fraction of a

---

[31] The telescope is described in *Sky & Telescope*, May, 1979, p. 491.

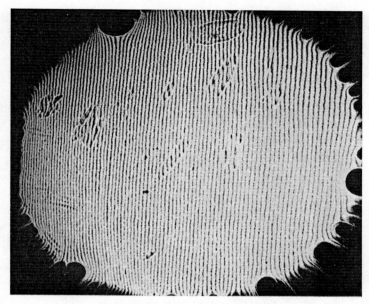

Figure 1.11. Interferogram of 15.2 cm (6 in) f/2 vacuum-formed mirror taken 20° off axis. The distortions at the rim of the mirror are caused by irregularities in the ring holding down the mylar. The oval shapes are water spots. Photo courtesy of Dr Peter Waddell, University of Strathclyde.

conventional mirror with the same aperture. Thus, the telescope structure can be made lighter too.

He has experimented with thin commercial grade window glass and found that the minor surface imperfections in the material, typically on the order of 0.01 mm (0.0004 in), are the limiting factor in mirror quality. In practice, he operates a manual vacuum control to optimize the focus. He notes that excessive vacuum can shatter the glass. While breaking your primary mirror can be traumatic, in his case flat window glass is cheap. I don't think I'd want to be standing next to the primary mirror, however, on some cold night while the focus is being adjusted.

The terms "active optics" and "adaptive optics" have grown in prominence recently and there is a bit of confusion as to their differentiation. While not all researchers agree, for purposes of this discussion "active optics" refers to the manipulation of the primary mirror or its segments over periods of several seconds to correct for mount and tube bending due to gravity and temperature gradients. "Adaptive optics" refers to the rapid bending of the light beam either at the primary mirror or along the light path to compensate for atmospheric seeing.[32] Early

[32] A summary of recent adaptive optics research is in *Sky & Telescope*, April, 1990, p. 361.

Figure 1.12. Vacuum-formed thin glass mirror. Photo courtesy of Guillermo Cock.

attempts at adaptive compensation for atmospheric seeing included a four-channel photomultiplier which measured the relative position of a guide star many times a second. This measurement could quantify the apparent lateral motion of the star due to seeing but did not attempt to measure the size of the star's image. The output of the four photometers was then fed to four speaker coils which moved a photographic plate holder back and forth in the image plane. Another similar arrangement measured the guide star position and simply closed a shutter for those fractions of a second when the star was displaced laterally, indicating poor seeing.

A more advanced seeing compensator was used by Horace Babcock at Mount Wilson Observatory. A rotating knife edge was placed at the focus of the telescope and a photomultiplier tube measured the intensity of the starlight. This information was then fed to a tip-tilt quartz plate which could displace the beam of the telescope, thus compensating for the wandering image of the star.[33]

Flexible glass mirrors have also been used as adaptive optics to distort the wavefront of light after entering the telescope. The Lockheed Corporation has developed a flexible mirror designed by Robert C. Smithson which tends to cancel out atmospheric seeing.[34] Sacramento Peak Observatory uses the device for solar work. In practice, a seeing monitor measures the wavefront distortion induced by the

---

[33] Private conversation with Horace Babcock. The rotating knife edge was a unique device consisting of a ball bearing running around the inside of a raceway, propelled by a varying magnetic field.

[34] *Sky & Telescope*, September, 1985, p. 219.

atmosphere. A computer then determines in near real time the amount of distortion present. Computer-driven mechanical actuators warp a small, thin flexible mirror at the image plane to compensate for the atmospheric distortions. Such technology is currently finding its way into the astronomical community via the "Star Wars" (Strategic Defense Initiative) research program.[35]

At present, only in radio telescopes where mechanical tolerances are much more relaxed, has active warping of a monolithic primary mirror been tried extensively. Mirror warping corrects the sag of the mirror under its own weight when pointed at different elevations. Rapid manipulation of the primary mirror surface to correct for atmospheric seeing has not been accomplished. For instance, the reflector of the 15 m (49 ft) aperture James Clerk Maxwell submillimeter radio telescope can be actively shaped by computer.[36] While this technology has been proposed for several optical telescopes, none are operational but several are under construction. The 3.6 m (142 in) New Technology Telescope (NTT) at the European Southern Observatory (ESO) is, at the time of writing, undergoing trials of its active optical system.[37] Active manipulation of several separate rigid mirror segments with respect to one another, as used on the Fred Lawrence Whipple Multiple Mirror Telescope on Mount Hopkins has almost become so common as to not deserve mention in a book on unique telescopes.

Gaseous mirrors

The final, and perhaps most exotic man-made mirror is one composed of gas. It has been proposed for a spaceborne radio telescope in the distant future.[38] Two lasers would be used to produce a standing wave field of light within a disk-shaped volume of hundreds of meters or perhaps several kilometers across. Gas molecules trapped in the field by photon pressure would act as a reflection hologram to radio waves and thus focus them onto a conventional radio telescope receiver. Such a structure may seem utterly fantastic to us but can you imagine trying to explain a rather conventional radio telescope array or a far UV telescope in space to, perhaps, Isaac Newton?

Planets, stars and galaxies used as optical elements

There are even more exotic materials for telescope optics, although they are not man-made. Astronomical objects themselves can be used as optical elements.[39] On 08 April 1976 astronomer James Elliot and his team from Cornell University were observing the occultation of the star Epsilon Geminorum by the planet Mars. They intended to measure the starlight as it passed through the upper reaches of the

---

[35] *Aviation Week & Space Technology*, 23 November, 1987, cover story and p. 80.

[36] *Sky & Telescope*, February, 1987, p. 153.   [37] *Sky & Telescope*, September, 1989, p. 250.

[38] *Sky & Telescope*, April, 1982, p. 338.   [39] *Sky & Telescope*, July, 1976, p. 25.

Martian atmosphere and thus determine the temperature and chemical composition of that atmosphere. They were flying aboard the NASA Kuiper Airborne Observatory which placed them directly on the center line of the occultation. After watching the star disappear, they waited for it to reappear. When the star was exactly behind the center of the planet, a bright flash was seen. Two theories emerged as to the cause of this effect. The first was that the planet formed the central ring of a Fresnel zone plate in which light was diffracted by the disk of the planet and formed a focus on Earth where it was observed. The second theory was that the tenuous gas in the upper Martian atmosphere formed an annular lens all around the planet and this focused the starlight. In either event, it has been shown that a planet can be used as a telescope objective.

A central flash of light has also been reported during an occultation by Neptune.[40] Similarly, during the 03 July 1989 occultation of 28 Sagittarius by Saturn's moon Titan, several astronomers in Europe recorded a broad central flash.[41] The flash was observed within a path several hundred kilometers wide and was probably caused by refraction in the tenuous atmosphere of Titan. Earlier that same night many American observers watched for a central flash as the star passed behind Saturn but none was seen, probably because the star did not cross the exact center of the disk as seen from Earth.

When the central flash was first observed in 1976, it was suggested that a gravitational lens had been formed. The gravitational lens is an effect predicted by Einstein's theory in which light passing near a large body is bent or deflected toward the body by gravity. The effect has been seen in starlight passing near the edge of the Sun as observed during a total solar eclipse. The stars appear to have moved slightly away from the Sun's position. This is an extremely difficult measurement to make and the deflection is minuscule for an object as massive as our Sun. The deflection of starlight passing near Mars would be even smaller due to the lesser mass of the planet and thus the Martian observation of 1976 was not an example of a gravitational lens.

Gravitational lenses have, however, been observed. Optical and radio astronomers have discovered several examples since the first seen in 1979 by radio methods.[42] In these extragalactic examples, the light from a very distant quasar is bent by an intervening galaxy composed of billions of solar masses of material. Depending upon the geometry of the situation, several separate images of the same quasar may appear all around the image of the galaxy. In some examples, the galaxy is faint and difficult to spot. In such a case, the clue is the discovery of two nearly

[40] *Sky & Telescope*, June, 1989, p. 608.
[41] Private communication with Dr David Dunham, International Occultation Timing Association. See also *Astronomy*, November, 1989, p. 52.
[42] *Sky & Telescope*, December, 1980, p. 486, November, 1983, p. 387, October, 1985, p. 319, November, 1986, p. 465, February, 1990, p. 127 and May, 1990, p. 471. See also *Astronomy*, October, 1987 p. 12 and May, 1990, p. 14.

identical quasars close to each other. In actuality, they are not separate quasars, but simply different images of the same one. A black hole could similarly bend light and produce a gravitational telescope.

Astronomers at Mount Hopkins have been monitoring a quasar by using an intervening galaxy as a part of the optical system. Since the quasar and the galaxy are not lined up exactly, the light path from the quasar past the galaxy to one of the images is several light years longer than the light path from the quasar to the other image. Since quasars can change their optical and spectral signature over a matter of weeks or months, monitoring the two images offers insight into the light path differences and thus into the structure and dynamics of a gravitational lens.

Most models of gravitational lenses have relied on the assumption that all of the mass of the lensing body is in one point. In reality, the mass of a galaxy is spread out among the billions of stars and perhaps even more dark matter. Recent models indicate that a galaxy would act like a very small (0.0001 arc second) multiple lens much like a fly's eye.[43]

Recently observers have reported seeing long arcs of light near galaxies. It is theorized that these arcs are indicative of the lensing galaxy being nearly aligned with the distant quasar. If the two were exactly aligned then a complete ring of light would be formed. As of this writing, a complete "Einstein Ring" has not been observed. I have full confidence that a little diligence at the telescope will soon render the previous sentence obsolete.

Classification by number of optical elements
Single-mirror systems

Having looked at optics from a standpoint of the mirror materials used, we will now consider complete optical systems in telescopes. A logical grouping is to classify the telescopes by the number of mirrors or lenses included. There are very few single-mirror systems. Usually, the image is reflected by either a Cassegrain secondary or a folding flat as in a Newtonian. Herschel's famous large wooden telescope, shown in Fig. 1.13, appears at first glance to be a Newtonian but, in actuality, the observer often stood at the end of the open tube and looked through an eyepiece attached to the inside wall of the tube which pointed straight back at the primary mirror. The system thus was a single-mirror off-axis reflector. This arrangement is often referred to as a Herschelian telescope.[44]

Herschel is, to amateur telescope makers, the philosophical equivalent of a patron saint. He made his most famous discovery, the planet Uranus, as an amateur astronomer and he built his own telescopes from scratch. If they didn't follow the conventional plan and trend of his times, then so be it. He usually broke new ground whenever he built or observed. The forty foot telescope was completed in 1789. It

---

[43] *Sky & Telescope*, May, 1988, p. 489.   [44] *Sky & Telescope*, June, 1986, p. 621.

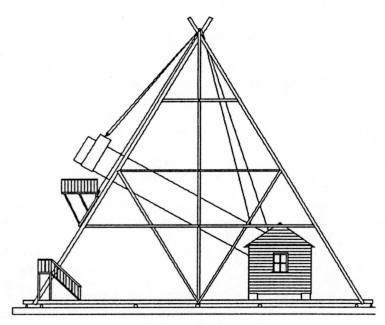

Figure 1.13. Herschel's forty foot telescope.

had a speculum metal mirror with a clear aperture of 1.22 m (48 in). One of the main reasons for eliminating the secondary mirror was the poor reflecting quality of mirrors in general. Speculum metal reflects only about 40% of the light. By using an off-axis eyepiece, nearly one visual magnitude more sensitivity could be obtained than with a Newtonian.[45]

The primary mirror, weighing over a ton, required frequent removal from the telescope and polishing in his optical shop to remove tarnish. Usually a mirror would last two or three months but, under some conditions, could tarnish within a week. Herschel made at least two mirrors for the forty foot telescope and kept one on the polishing bench while the other was being used. Although the telescope could be moved to any position in the celestial sphere, it was most often used as a transit instrument for Herschel's surveys of the sky. The small building housed the recorder, usually Caroline Herschel, who noted each object which William described as the sky moved past his eyepiece.

While the telescope was the largest in the world at the time and its construction certainly advanced the state of the art in technology, it was not used as extensively as his similar twenty foot telescope which had better resolution and was much easier to use. Although optical technology had reached the point where meter-class

---

[45] The development of this telescope is detailed in *The Telescope*, Louis Bell, Dover Publications, 1981 edition, p. 33, and *The History of the Telescope*, Henry C. King, Dover Publications, 1955, p. 129.

instruments could be built, mechanical fabrication of large bearings, gears and drive systems was not possible at the time. Thus, telescopes of the late 1700s were generally positioned with ropes, pulleys and muscle power. Well-balanced motor-driven equatorial mounts would come in the following century.

During the 1800s Herschellian eyepieces were designed into telescopes made for commercial sale to the public. One obvious advantage is the decreased cost since a secondary mirror isn't needed. Amasa Holcomb produced many telescopes of this type and standardized the design. From 1828 to 1842 Holcomb was the only commercial telescope manufacturer in the United States and some of his telescopes survive to this day.[46]

A defect of the Herschellian telescope is that it places the observer near the light path and his body heat may disrupt the seeing. This has, at times, been alleviated by placing a small folding flat near the end of the telescope to one side of the light path. The light is then directed out through the side of the tube to an eyepiece which resembles a Newtonian design, as shown in Fig. 1.14. The advantage of this arrangement is that it moves the observer away from the light path while retaining the desirable feature of not having the secondary mirror obscure part of the incoming light, as in a Newtonian.[47] Since the secondary mirror is out of the light path, diffraction caused by the smaller mirror or its supports is not introduced into the system. The addition of this mirror, however, makes the instrument a two-mirror telescope with all of the light losses of a second mirror.

Single-mirror telescopes have been used in more modern systems, as illustrated in Fig. 1.15. The United States Air Force Avionics Laboratory Observatory near Wright-Patterson Air Force Base, Ohio, observed with a 63.5 cm (25 in) single-mirror satellite-tracking telescope. The telescope was designed and made by Group 128, Inc. and used a TV camera at the prime focus of the f/3 mirror. At times a corrector lens was used just in front of the camera to flatten the fast, wide field image plane. The telescope also features several design innovations such as a one-arm fork and a tapered stressed-skin tube, fabricated much like a monocoque aircraft fuselage. Light weight and stiffness criteria were imposed on the telescope design because it was required to slew rapidly from point to point in the sky, stop, search for a few seconds and then slew rapidly to the next point. The telescope could be operated remotely by either manual or computer control. Generally, during observations no astronomers were in the dome. They sat at a console away from the cold night air and observed in comfort. The telescope is also shown in Fig. 9.5.

Two-mirror telescopes are the commonest variety. Typical examples are the classical Cassegrain and Newtonian systems. There are, however, many unusual variations upon these themes. The simplest approach to a two-mirror telescope is the rail type. It is the no frills, bare bones engineering approach to telescope design. This is basically an optical bench bolted to a mount. It has the advantage over telescopes

---

[46] A typical Herschellian telescope with a Holcomb mount is shown in *Sky & Telescope*, June, 1986, p. 620.   [47] *Sky & Telescope*, September, 1958, p. 589.

Figure 1.14. Herschelian telescope with Newtonian folding mirror near the front aperture. Photo courtesy of Oscar Knab.

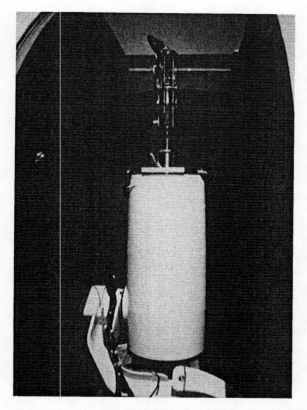

Figure 1.15. 63.5 cm (25 in) f/3 telescope used at prime focus for search and detection of artificial Earth satellites

with tubes in that it can be folded and transported in a much smaller volume.[48] The telescope built by Joe Perry of Las Vegas, Nevada is an excellent example. The instrument, shown in Fig. 1.16, has several innovative design features which make the system more useable. Foremost among these is the ring at the eyepiece end which allows the eyepiece and secondary assembly to rotate about the optical axis. This is analogous to a classical Newtonian telescope with a rotating tube which allows the eyepiece to be positioned at a comfortable location no matter where the telescope is pointing.

[48] There is an old and unresolved argument in telescope design circles about disassembling a telescope for transport. Proponents claim a disassembled telescope takes up less space and thus allows transport of larger telescopes. Opponents claim that the time spent realigning the optics does not make disassembly worth it. On the other hand, careful design with alignment pins can speed up assembly and alignment. Then there are those, like myself, who don't disassemble the optics yet check alignment anyway. This discussion, oft repeated, easily drifts away from operational and engineering rationale into the philosophy of observing methods.

Figure 1.16. Rail-mounted Newtonian. The rail is the square steel beam which supports all of the optical components. Note that the eyepiece and secondary holder rotate around the optical axis for comfortable positioning of the eyepiece. Photo courtesy of Joe Perry.

An advantage to open-tube designs is that they stabilize in temperature much more rapidly than telescopes with tubes. The open-air light path, however, makes them more susceptible to the problems of thermal currents generated by observers and other heat sources. While such a telescope is very portable, the lack of baffling can cause stray light to enter the optical path. For instance, the eyepiece can see any distant light sources in the direction of the secondary mirror. For this reason, many open-tube Newtonians have a solid black light baffle across the tube from the eyepiece.

A second "minimum structure" telescope is a tubeless Cassegrain system, not to be confused with the Dilworth telescope which has auxiliary field correction lenses

Figure 1.17. Tubeless Cassegrain telescope at Mount Hopkins, Arizona. Photo courtesy of
Russell Genet, Fairborn Observatory.

within the baffle tube and is described elsewhere. A Cassegrain telescope without an external tube must be carefully designed to prevent off-axis stray light from entering the optical system. This usually means a large central baffle tube as shown in Fig. 1.17. Without a conventional outer tube, the spider becomes nearly a tripod. This particular telescope was designed for photoelectric photometry for the Smithsonian Astrophysical Observatory. It does not have provision for an eyepiece so there is no need for a large working space behind the primary mirror cell. Thus, the compact horseshoe mount design can be used.

The instrument, operated by the Automated Photoelectric Telescope Service (APT), is also unusual in that it is fully automated (computer driven) and operates unattended on clear nights at Mount Hopkins, Arizona. The computer knows when the Sun sets and it checks the weather for clouds, rain or snow. If the skies are clear, the computer opens the roof and starts on a preprogrammed sequence of photometric measurements. It continues measurements until dawn or weather intervene. The sequence and priority of observations can be altered from a remote computer via telephone communication. It has been quipped that future astronomers won't stay up all night but their telescopes will. Then the telescope computers will call the astronomers at a decent hour the next morning and tell them how beautiful the night was. I wouldn't bet on it, however. Astronomers like to get out under the sky in the shadow of the telescope. I know of one designer of automated telescope systems who lets the telescopes churn away on the mountain but he also drags his 20.32 cm (8 in) telescope out in the back yard often to do "real" astronomy.

A third two-mirror telescope actually uses a solar filter as a mirror. In this design, adapted by many Californian amateur solar observers, a full aperture aluminum or Inconel[R]-coated flat glass solar filter is tilted 45° with respect to the optical path at the front of the telescope.[49] The Sun's rays pass through the filter to a rather conventional primary mirror and are reflected back toward the front of the tube. The backside of the shiny solar filter then acts as a tilted secondary mirror, directing the rays to a conventional Newtonian eyepiece holder. Theoretically, such an arrangement would not work well for stellar observations even if the Inconel[R] coating were removed because the eyepiece can see quite a bit of the ground surrounding the telescope and the system is not well baffled against off-axis stray light. Since the Inconel[R] coating of the solar filter blocks most such stray light and there are no bright sources other than the Sun, the system works well for solar observations. Generally, the primary mirror in such telescopes is left uncoated.[50] A similar concept using an uncoated primary and a welder's glass diagonal/filter was seen at the Riverside Telescope Makers Conference, in 1989.[51]

A smaller version of this idea using a welder's glass was made by Steve O'Dwyer of the East Valley Astronomy Club. Since the welder's glass has a very low reflectivity when used as a secondary mirror, the primary mirror is aluminum coated. The simple cardboard tube and the method of holding the filter on with rubber bands appeal to one's sense of efficiency in use of materials as shown in Fig. 1.18. Note that the telescope does not have a finder scope. The method of aligning the scope on the Sun is to move the scope around until the tube is entirely in the shadow of the welder's glass.

The classical Schmidt camera, while a single-mirror system, has been modified occasionally to a two-mirror system. This is because the image plane of a Schmidt camera falls about half-way between the corrector plate and the primary mirror. If you are going to mount a photographic film or plate holder in this position, then there shouldn't be too much trouble. On the other hand, if you need a bulky TV camera or photoelectric sensor then the camera body will often block most of the aperture. A solution to this is to bend the optical path with a second mirror and allow the image plane to protrude through the folding mirror as shown in Fig. 1.19. An example of this is a United States Air Force prototype satellite-tracking camera at Rimfire Site, White Sands Missile Range, New Mexico. Fig. 1.20 shows a Celestron 35.56 cm (14 in) Schmidt camera which has been modified with a folding flat to allow a large TV camera to be used. The folded Schmidt, on the right side of the mount, and

---

[49] An early reference to this type of telescope is in *Telescope Making*, No. 8, Summer, 1980, p. 40. The design, pioneered by John Dobson, is shown in *Telescope Making*, No. 20, Summer/Fall, 1983, p. 21.

[50] *Sky & Telescope*, August, 1989, p. 207.

[51] The telescope, by Tom Mathews, is shown in the *Proceedings of the Riverside Telescope Makers Conference*, May 26–9, 1989 p. 30.

Figure 1.18. Solar telescope using welder's glass as both a filter and as the secondary mirror. Photograph courtesy of Steve O'Dwyer.

the larger main telescope both point at the same position in the sky. The large dark circular structure on the front of the Schmidt is the primary mirror housing. The TV camera mounts on the rear of the Schmidt assembly and is the same size as the camera on the bottom of the larger telescope.

Since the image plane of a Schmidt camera is a curved surface, either a field-flattening lens is required, as shown in Fig. 1.19 or the TV camera fiber optic faceplate must be ground to a corresponding curve. A similar off-axis Schmidt camera produced for NASA had the unique feature that the folding flat also served as the corrector plate.[52] The reason for an all-reflecting Schmidt was to observe outside the usual visible wavelengths and into the ultraviolet, where refracting corrector plates would have had chromatic aberration problems.

One advantage of the bent Schmidt is that it minimizes blockage of the aperture by detectors, secondary mirrors and mechanical structure. There is, however, still a portion of the aperture which is taken up by the hole in the folding mirror through

[52] *Sky & Telescope*, April, 1967, pp. 201 and 204.

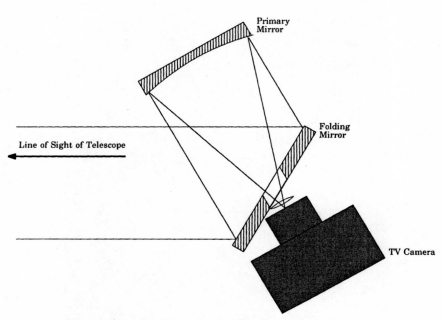

Figure 1.19. Optical path of bent Schmidt telescope.

which the camera looks. Even this small loss can be eliminated in an off-axis Cassegrain, shown in Fig. 1.21.

The off-axis Cassegrain telescope requires a parabolic surface just as in a conventional Cassegrain. The center of curvature of the parabola, however, is at the edge of the mirror closest to the secondary and not at the center of the primary. Think of it as a classical donut-shaped Cassegrain mirror in which most of the glass except a round piece stretching from the original Cassegrain hole to the edge has been chipped away. This would be a very difficult figure to grind from scratch but for a radio telescope requiring less surface smoothness, it is relatively easy. The telescope in Fig. 1.21 has been proposed by NRAO to replace the late One Hundred Meter Transit Instrument which collapsed at Green Bank, West Virginia. In order to gain perspective on the size of the instrument, note the instrumentation house and elevator. The mirror itself is about the size of a football field.

Two-element optical systems include Schmidt cameras which have one mirror and one lens, although it is usually a weak aspheric correcting lens. An advantage of a Schmidt system is its relatively fast focal ratio. In the period before World War II, D. O. Hendrix of Mount Wilson Observatory experimented with a solid Schmidt and thick mirror Schmidt (Fig. 1.22). If the material in a solid Schmidt has an index of refraction of n then theoretically the system can be made n^2 times faster than a similar thin mirror Schmidt.[53] A solid Schmidt using diamond optics could have an effective

[53] *The History of the Telescope*, Henry C. King, Dover Publications, 1955, pp. 362–3.

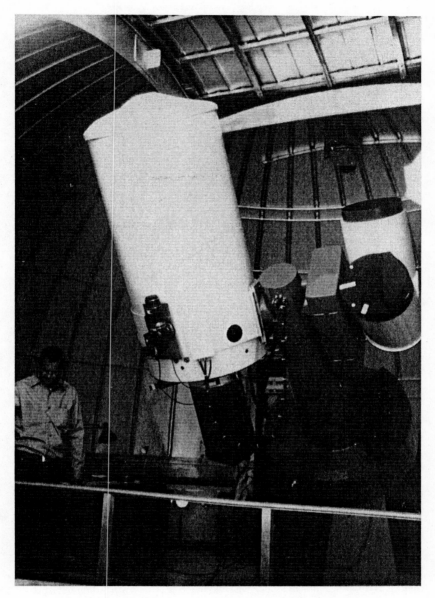

Figure 1.20. Folded 35.56 cm (14 in) Schmidt telescope for satellite detection.

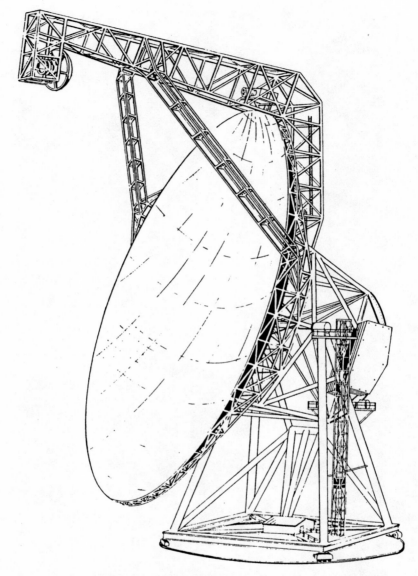

Figure 1.21. Proposed off-axis Cassegrain radio telescope. Drawing courtesy of National
Radio Astronomy Observatory (NRAO).

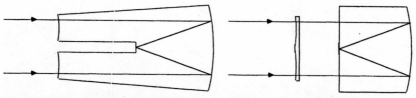

Figure 1.22. Solid and thick mirror Schmidts. The curvature of the corrector plates has been exaggerated.

aperture ratio of f/0.2, providing the small matter of materials expense could be solved. Hendrix constructed a thick glass mirror Schmidt with an aperture of 5.4 cm (2.125 in) and f ratio of 0.62. As with most on-axis systems, the film holder obscures a portion of the aperture and decreases the amount of light which can be received.

Off-axis two-mirror systems are generally designed to avoid obscuration by a secondary mirror as is commonly found in Cassegrain and Newtonian systems. Such secondary obscuration often adds problems of diffraction from the secondary mirror supports. An excellent example of an off-axis reflector is Joseph Pawlick's 21 cm (8.25 in) telescope.[54] A 25.4 cm (10 in) flat mirror sends the beam to an off-axis spherical primary mirror and then back to eyepiece which is mounted in a hole in the flat. Usually, a mirror used off axis in this manner introduces astigmatism. The 5.4 m (213 in) long focal length of this instrument, however, minimizes that effect. The long focal length also allows the mirror to be spherical rather than the conventional paraboloid. While the ideal mirror for this type of instrument would be an off-axis paraboloid, such a surface is difficult to grind. For this application, a spherical mirror suffices. Due to the length of the instrument plus the fact that there is a heavy mirror at both ends of the telescope, quite a bit of trusswork is involved in keeping the mirrors aligned and stabilized. The whole assembly is susceptible to vibrations. Indeed, it took Mr Pawlick four years to perfect the mechanical arrangement of the instrument and make it useable.

Grinding an off-axis paraboloid, a daunting task, can be accomplished easily if the mirror blank is thin. Several such mirrors have been figured with the mirror face up on the grinder. While supporting the mirror at its center, weights are hung from the back of the mirror, deforming it slightly. The mirror is then ground to a sphere, an easy task. When the weights are removed, the mirror then flexes back to its original shape and the spherical front surface will be distorted into an off-axis paraboloid. The trick is in calculating the placement and mass of the weights. This technique is similar to the vacuum chucks used to produce Schmidt corrector plates.

One way to get around the requirement that an off-axis paraboloid be ground is to use a thin mirror and warp it slightly by stressing the mirror mounting. Arthur Leonard's Yolo telescope (named for the area where he lives north of San Francisco,

[54] The telescope is pictured in *Sky & Telescope*, October, 1966, p. 231.

Figure 1.23. Yolo telescope shown at the Riverside Telescope Makers Conference in 1981.

California) uses this principle. The telescope, shown in Fig. 1.23, uses a 15.24 cm (6 in) concave spherical primary and a 10.8 cm (4.25 in) concave spherical secondary located off axis in the tube. In order to correct for the astigmatism the secondary is warped into a "potato chip" shape (in Metric countries this is referred to as a "potato crisp").[55] So far, all instruments of the Yolo type have relatively long focal lengths.

The schiefspiegler

The schiefspiegler telescope (German for oblique reflector) is a classic off-axis reflecting telescope. Several versions of its design history have been presented in the literature.[56] The type was originally called a brachyt and was invented in Vienna about 1876. Other sources place the invention by Johann Zahn in 1865.[57] Anton Kutter of Germany revived it in the 1950s and gave it the present name. Both two and three mirror variations have been produced as shown in Fig. 1.24. My children call the schiefspiegler a sheep sprinkler.

The main characteristic, as shown in the finder scope in Fig. 1.25, is a spherical

[55] *Sky & Telescope*, August, 1979, p. 113. See also *Sky & Telescope*, August, 1988, p. 198 for Jose Sasian's Yolo. See also *Telescope Making*, No. 4, Summer, 1979 p. 30.
[56] *Sky & Telescope*, April, 1965, p. 248 and May, 1985, p. 461. The word brachyt in German means broken or bent.   [57] *Sky & Telescope*, August, 1958, p. 529.

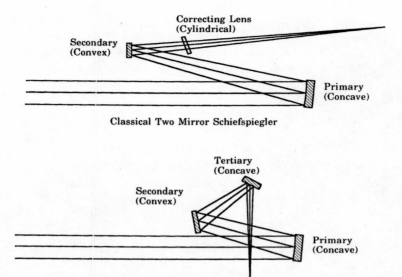

Figure 1.24. Common schiefspiegler optical paths.

primary mirror housed in the short tube under the eyepiece. The light then passes through an opening in the underside of the longer tube and strikes the convex secondary which has the same radius of curvature as the primary. Light then passes down the small tube to the eyepiece. In order to help understand the relationships of the optical surfaces, Oscar Knab's drawing of a similar schiefspiegler is shown in Fig. 1.26.[58]

In some schiefspiegler versions a weak cylindrical lens is placed between the secondary and the focal plane as shown in Fig. 1.24. This optical concept is designed to reduce coma and results in a flat image plane. It should be noted that the telescope's line of sight is at a slight angle to the direction in which the longer tube points. This often results in frustration as observers make coarse alignments of the tube with astronomical objects of interest and then look through the telescope, only to find that they are several degrees away from their intended position. This slight inconvenience is more than made up for in the superb lunar and planetary views afforded by the type.

An interesting mechanical implementation was developed by F. Salomon of Haifa, Israel in which the mass of the primary mirror and housing were used as a counterweight, thereby decreasing the overall weight of the telescope.[59] Schiefs-

---

[58] Oscar Knab outlined the making of a schiefspiegler telescope in an article in *Telescope Making*, No. 1, Fall, 1978, p. 4. [59] Salomon's telescope is shown in *Sky & Telescope*, August, 1958, p. 529.

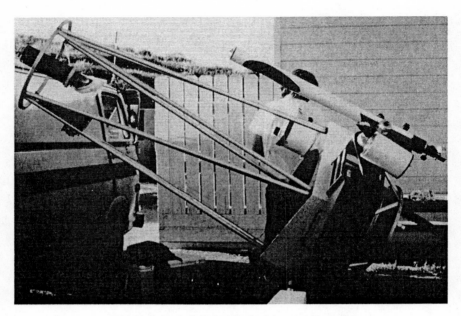

Figure 1.25. Schiefspiegler finder telescope by Kevin Medlock of the San Jose Astronomical Association. Note that neither the long tube nor the primary mirror holder point exactly along the main telescope's line of sight.

pieglers generally are high f number systems and as such are not often used for deep sky observing. There seems to be no limit on the size of the system and they have been made as small as a 60 mm (2.4 in) f/24 system[60] and as large as 30 cm (11.8 in) in aperture.

Jose Sasian of Tucson, Arizona, designed an alternate off-axis two-mirror system, the unobstructed Newtonian[61]. The telescope diagram, shown in Fig. 1.27, uses a paraboloidal primary and a small concave secondary which cancels the aberrations introduced by the tilted primary. The secondary is a concave oblate ellipsoid with different radii of curvature in different directions around its rim, much like the surface of an egg. While this may seem to be an impossible surface to grind, the differences in radii are small and can be added to the mirror during final polishing. Jose has done extensive work in optimizing the design using a small computer. The design could be considered as a variant of the Yolo telescope. Jose's telescope was first shown at the

---

[60] *Sky & Telescope*, December, 1967, p. 403.

[61] *Telescope Making*, No. 38, Fall, 1989, p. 4. See also *Proceedings of the Riverside Telescope Makers Conference*, May 26–9, 1989, p. 1. The general design of off-axis unobstructed reflectors is covered in an article by Dr Sasian in *Optical Engineering*, Vol. 29, No. 1, January, 1990, p. 1. See also *Telescope Making*, No. 38, Fall, 1989, p. 10, and *Sky & Telescope*, December, 1989, p. 594.

Riverside Telescope Makers Conference in 1989. A photograph of the instrument is shown in Fig. 4.10. While the telescope gives good views for a moderately sized instrument, the position of the primary mirror near the ground makes it susceptible to damage by the observer's feet and allows dust to collect on the primary. In addition, any flashlights shown at the ground will brighten the field of view. A lightweight tube structure enclosing the optics should cure these minor problems.

In order to eliminate the problems associated with generating and testing the concave oblate ellipsoid secondary mirror, Ed Jones of Cincinnati, Ohio placed a small meniscus lens just in front of the flat secondary.[62] While the meniscus lens uses all-spherical surfaces, it has a slight wedge between the front and back curves. Since the design includes an enclosed tube it most resembles a Herschellian telescope and Ed has labeled it an unobstructed catadioptric Herschellian. A similar telescope was made much earlier by the British amateur J. S. Hindle.[63]

Three-mirror systems

There are several reasons to add a third mirror to a telescope's optical system. Often the additional mirror is not a flat and is used to correct residual aberrations in the other optical elements. In some designs a third mirror is added to fold the optical system into a shorter package or bring the image out at a convenient place on the tube. The classical three-mirror telescope, however, is the tri-schiefspiegler, shown in Fig. 1.24. This design is an outgrowth of the original schiefspiegler and it incorporates the same concept of an unobstructed reflector.[64]

The most obvious advantage of this design is that it folds a relatively long telescope into a more compact box. The disadvantage is that it requires grinding an extra curved surface, although the third mirror is only a weak sphere. The type has been produced in several versions such as that made by Oscar Knab, shown in Fig. 1.28. The unobstructed high f number optical system works best on planets and the Moon, yielding crisp images with a minimum of diffraction. Not all tri-schiefspieglers have a curved third mirror, however. Some are merely a conventional schiefspiegler with a folding flat which makes the optical layout more compact. If the third mirror is a flat then the design works best only if the focal length is relatively long.

Another tri-schiefspiegler telescope, built by A. L. Woods of Kirkwood, Missouri, is shown in Fig. 1.29. Due to the folding of the optical system a compact design is possible, which allows the telescope to swing through the forks. The 15.24 cm (6 in)

[62] See *Telescope Making*, No. 38, Fall, 1989, p. 10 and *Sky & Telescope*, December, 1989, p. 594.

[63] *Journal of the Optical Society of America*, Vol. 34, 1944, p. 270.

[64] General articles on tri-schiefspieglers by Richard Buchroeder are in *Telescope Making*, No. 28, Fall, 1986, p. 33, and *Sky & Telescope*, December, 1969, p. 418. A pair of articles by Anton Kutter on design and fabrication of the type are in *Sky & Telescope*, January, 1975, p. 47 and February, 1975, p. 115.

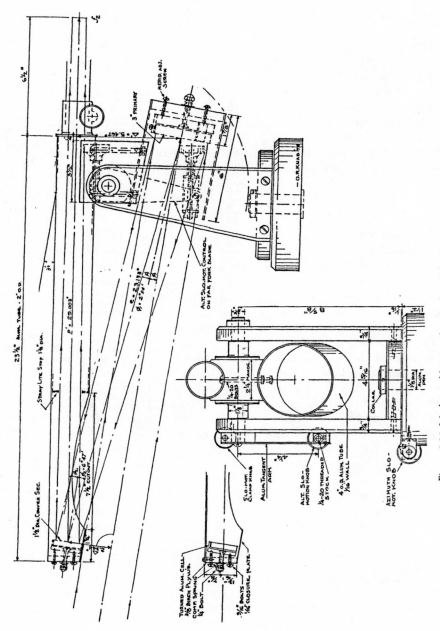

Figure 1.26. Mechanical layout of a schiefspiegler. Drawing by Oscar Knab.

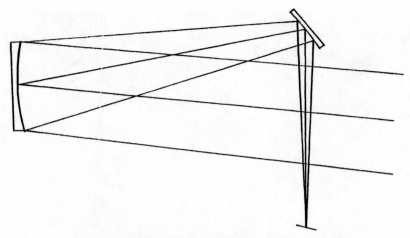

Figure 1.27. Unobstructed Newtonian telescope by Jose Sasian.

aperture instrument also features a very stable mount with short, massive forks.[65] The mount has an interesting cast sector RA (right ascension) wheel as discussed in chapter 4.

Systems with three mirrors sometimes use folding flats added to a conventional Cassegrain or Newtonian optical layout. The usual reason is to bend the optical path so that the image falls at a desired position. A typical example is the tertiary mirror used in the 30.5 cm (12 in) telescope designed by Max Bray of Phoenix, Arizona, in the Funds for Astronomical Research (FAR) telescope. It has an interesting el-el mount as shown in Fig. 2.9. The telescope is intended to be a quarter scale model of a spectroscopic telescope. The spectroscope is much too large to be mounted on the telescope so it is placed at the Southern end of the polar axle. The optical system is composed of a primary and secondary in the Cassegrain style plus the folding mirror. The tertiary mirror directs light from the Cassegrain secondary out through a hole in the South axle, shown in Fig. 1.30. The telescope is probably more of a coudé than anything else and resembles many of the larger coudé systems on professional telescopes except that the spectroscope is mounted horizontally instead of pointing up the polar axle.

One of the problems associated with such a system is that the third mirror must

[65] The construction and collimation of tri-schiefspieglers is explained in a pair of articles by A. L. Woods in *Telescope Making*, No. 16, Summer, 1982, pp. 10 and 18. The telescope described also appears in *Telescope Making*, No. 30, Summer, 1987, p. 10. An earlier version of the telescope is in *Telescope Making*, No. 10, Winter, 1980/1981, p. 29. A similar pair of articles details the construction of a Buchroeder tri-schiefspiegler by Richard J. Wessling and a conventional tri-schiefspiegler by Steven W. Johnson in *Telescope Making*, No. 28, Fall, 1986, pp. 32 and 44 respectively.

Figure 1.28. Tri-schiefspiegler telescope. Photo courtesy of Oscar Knab.

Figure 1.29. 15.24 cm (6 in) Kutter tri-schiefspiegler. Photo courtesy of A. L. Woods.

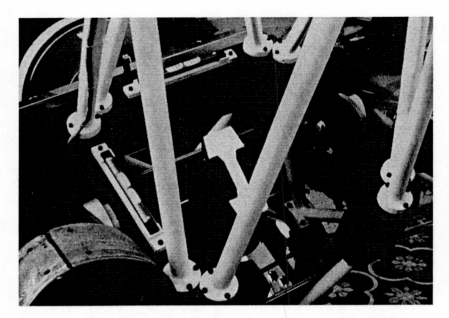

Figure 1.30. FAR telescope designed by Max Bray with a tertiary mirror to direct the beam out through the South polar axle to a large spectroscope. A photo of the entire telescope including the el-el mount is shown in Fig. 2.9.

turn exactly half of the rotation of the inner axle which corresponds to the declination shaft. While the tertiary mirror position is controlled by a 2:1 gear train from the declination drive, the designer recommends that stepping motor control of the position would probably be more efficient. The reason for this is that the 2:1 gear train must have no backlash or the image will shift across the image plane when the inner axle is moved. The spectroscope design is very sensitive to the image shift problem. The telescope was installed in 1985 on campus at the University of Chicago for use as a teaching instrument.[66] It has a computer control system which allows it to be placed at any latitude. Indeed, if it were placed at the equator then it would be a conventional equatorial mount. The computer converts celestial coordinates to the unique el-el mount coordinates and commands stepping motors to drive the telescope at the correct rate in two axes in order to track the stars. It should be noted that this results in an apparent rotation of the field of view as the telescope tracks. Since the system was designed for spectroscopy, however, this is not considered critical.

The bent Cassegrain telescope was first suggested in 1876 by Blacklock[67]

[66] The installation is described in *Sky & Telescope*, November, 1988, p. 567.
[67] This information has occasionally been disputed. For a discussion of the topic see *Sky & Telescope*, May, 1944, p. 9.

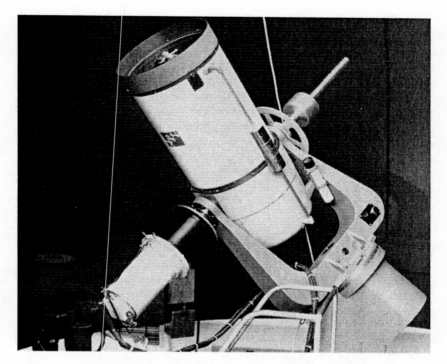

Figure 1.31. Bent Cassegrain telescope. Celestron 40.6 cm (16 in) commercial telescope modified to bent Cassegrain configuration for addition of a large TV camera. Note the counterweight opposite the TV camera.

although the basic idea was elaborated by Nasmyth in the 1840s.[68] The Nasmyth telescope, usually associated with alt-az mounts, is listed in chapter 6. The idea of bringing the image out through the declination axle has also been applied to Cassegrains on equatorial mounts. A typical bent Cassegrain system, as shown in Fig. 1.31, was used by the United States Air Force for television camera tests to develop satellite tracking equipment. The 40.6 cm (16 in) aperture Celestron telescope was modified from the Cassegrain to the bent Cassegrain configuration in order to allow the large TV camera to be mounted. The camera was so long that if it were mounted at the Cassegrain focus, it would not pass through the forks and would also strike the mount in many positions. Since the camera was fairly heavy, a counterweight was required on the opposing declination axle, thus giving the system a cross-shaped appearance. This is the same telescope as that pictured in Fig. 5.2 after modifications.

[68] *Astronomy*, Fred Hoyle, Crescent Books, 1962, p. 57.

A modification of the bent or broken Newtonian is Ed Danilovicz's 31.1 cm (12.25 in) f/6.4 telescope with a flat secondary set at a slight angle to the optical axis.[69] Originally built as a classical Newtonian, the secondary was modified so that the eyepiece position did not require ladder access. During the modification, a curved secondary mirror holder was added in order to minimize diffraction effects. This technique is also discussed under the section on eliminating diffraction spikes. In Fig. 1.32 Mr Danilovicz is shown looking through the main eyepiece which is near the small tertiary mirror. The design is unique in that the eyepiece holder and tertiary assembly are hinged to swing out of the way, allowing the light to travel farther downward to a camera mount. This telescope is a classical example of ingenuity in use of materials. The mirror enclosure is a recycled searchlight case. There are numerous plumbing fixtures visible on the telescope and the control system is a collection of "available parts".

Wide field optics

While most telescope designs concentrate on high magnification with an implied narrow field of view, there are many applications in which a wider field of view is required. This implies a relatively short focal length and the problem is not one of producing a large enough objective mirror or lens, but of making an aberration-free instrument of low f number. As an example, the Harvard Observatory wide field patrol cameras have a focal length of only 30.48 cm (12 in) which doesn't seem too difficult to manufacture. The trick is to make sure the lens produces a high resolution field across a photographic plate which measures 20.32 cm by 25.4 cm (8 in by 10 in).[70] The cameras are used to make nightly records of selected areas of the sky. Each plate records the brightness of thousands of stars. Astronomers trying to determine if a given star is variable can go back through years of plates to examine it over long intervals. While the method is not as accurate in measuring star brightness as photoelectric photometry, the decades of photographic plates at Harvard and other similarly equipped observatories are a rich treasure of data on millions of stars.

Wide field cameras and telescopes are also useful in the study of meteors. Since the astronomer has little idea of exactly where a meteor will appear, he must monitor a large section of sky to get any data at all. James G. Baker of the Perkin–Elmer Company designed a meteor camera with an effective aperture of 31.1 cm (12.25 in) and an effective f number of 0.82 as shown in Fig. 1.33. This 55° field of view system employs two nearly concentric meniscus lenses of about a half meter diameter. Between the meniscus lenses is a two-element achromatizing corrector plate. The rays actually pass through the second meniscus lens twice on the way to the film plane. The film plane is highly curved, with a radius of only 20.32 cm (8 in). This is

---

[69] *Sky & Telescope*, April, 1976, p. 278.

[70] The patrol cameras are described in *Sky & Telescope*, April, 1965, p. 202. A similar camera is shown in the June, 1982 issue on p. 556.

Figure 1.32. Bent Newtonian telescope. Photo courtesy of Ed Danilovicz.

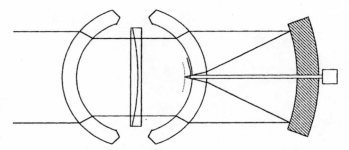

Figure 1.33. Baker Super–Schmidt camera optics.

too small a radius for bending glass plates and even special film had to be molded with heat and held in place with a vacuum.

The 58.4 cm (23 in) aperture mirror is spherical and is pierced by a rotating rod that controls a propeller-shaped shutter just in front of the curved film plane. As the rotating shutter alternately covers and uncovers each section of film, a meteor streak produces a dashed line on the negative. The length of the dashes indicates the velocity of the meteor. Several of these cameras were built and two of them were placed at the Harvard Observatory meteor patrol station near Las Cruces, New Mexico. The cameras, placed about 30 km (18 mi) apart could record the passage of a meteor from two different locations, thereby allowing triangulation to determine the height and direction of the meteor.[71]

A similar design by Baker was used for keeping track of artificial Earth satellites. Originally conceived for the Moonwatch Program in the late 1950s, a dozen of the systems were built. Baker–Nunn telescopes, shown in Fig. 1.34, were used by the United States Air Force for years to observe satellites, spent booster rockets, shrouds and all matter of space junk.[72] Several of the telescopes were scattered around the world to provide complete coverage of all possible orbits.

The three-element corrector of the camera was 50.8 cm (20 in) in diameter and the primary mirror was 78.75 cm (31 in) across. Although the f/1 optics were very fast, there was considerable obscuration of the curved film plane by internal components. The film holder and remotely controlled shutter were supported on rather thick spiders which were hollow in order to allow the film to pass through. Film was stored in supply and take-up reels on either side of the optical system.

The optics were carried by a motorized three-axis mount which allowed the telescope to track a predicted satellite position as it moved across the sky. A photo of the mount with a different set of optics is shown in Fig. 2.13. By tracking the satellite

---

[71] The Super-Schmidt camera is described in *The History of the Telescope*, Henry C. King, Dover Publications, 1955, p. 367. It is also pictured in *Sky & Telescope*, February, 1957, p. 168.

[72] Baker–Nunn cameras are described in *Sky & Telescope*, December, 1966, p. 338, November, 1968, p. 285, and September, 1975, p. 160.

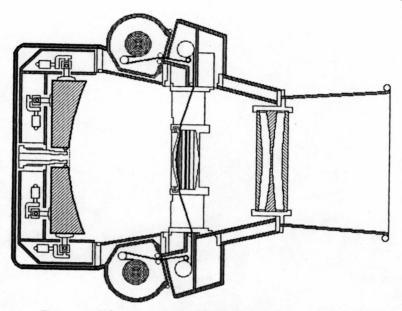

Figure 1.34. Baker–Nunn camera. The three-axis mount is not shown.

and not the stars, much fainter satellites could be observed. This is analogous to tracking a comet nucleus during a time exposure. The images of the stars form streaks but, in this case, the stars are only of secondary interest. The stars were not totally ignored, however, since the satellite's position was measured with respect to the background stars.

Typically, many exposures would be made in one night and the film would be processed and reduced the following day. Thus, the system determined where the satellites were during the previous night but could not determine real-time information on satellite positions. This system was used until the more modern real-time TV based satellite-tracking systems were developed in the late 1970s. The Baker–Nunn cameras were also used for classical astronomical wide field photography, meteor photography and geodesy.

The widest field usually attempted is a full 180° hemisphere. Such an all-sky view is useful in studies of meteors and auroras. Dr Michael Gadsden of Aberdeen University has developed an all-sky camera using two convex mirrors, as shown in Fig. 1.35.[73] The box below the optical system contains a film camera used to record the meteors and a timer system which controls exposures. The system was originally developed for the International Geophysical Year (1957–8) and has operated nearly continuously since then. The reason for employing convex mirrors is to avoid the

[73] *Journal of the British Astronomical Association*, Vol.88, No. 6, October, 1978, p. 570.

Figure 1.35. All-sky meteor camera at Aberdeen University. Photograph is reproduced by permission of Dr Michael Gadsden, who is the owner of the copyright.

common refractor elements used in wide field or fish-eye lenses. Such lenses typically have an f number of about 5 or 6. The all-reflecting system has an f number of about 2. Thus, this system is more sensitive to faint meteors but with its lower f number it will reach the background sky fog limit with a shorter exposure. Increased sensitivity is offset by the obscuration of a small section of sky at the zenith caused by the secondary mirror suspended from the top of the clear weather dome. On the other hand, large aperture all-reflecting optics are much less expensive than refractor lenses of similar performance.

Both reflector and refractor all-sky systems have extreme distortion at the edge of the frame since the hemispherical three-dimensional sky is imaged onto a flat, round negative. This can be troublesome if the positional track of a meteor is to be plotted from the negative image. The distortion can, however, be measured and positional corrections applied during data reduction.

The hub cap all-sky camera is one of those ideas which seem to get reinvented about every two decades.[74] It is basically a simple device consisting of a convex mirror pointing upward with a conventional 35 mm camera suspended over it.[75] The system differs from Dr Gadsden's all-sky camera in that the secondary has been replaced by a camera at the prime focus of the mirror. The image of the camera appears at the center of the picture but it usually occupies a very small area on the negative. In addition, the camera must be supported so it is often placed on a tripod whose legs span the primary mirror. The images of the legs then cut the picture into three sectors. Many conventional meteor-patrol cameras have been built along these general lines.

Several approaches have been taken to eliminate the images of the tripod legs. Jeff Charles and Chris Schur of the Saguaro Astronomy Club have done extensive work on the problem. The most common design uses a dome similar to the one used by Dr Gadsden to support the camera. A second approach is to run a single strong rod from the center of the mirror to the center of the front lens element on the camera, as shown in Fig. 1.36.[76] In actuality, the camera's front lens element isn't pierced by the rod. A plane sheet of glass with a hole in it is mounted in a filter holder on the front of the camera. This approach works only for small cameras and care must be taken to avoid vibrations in the rod. With optics such as these, Charles and Schur have photographed the Milky Way, a few bright meteors and the Zodiacal Light.

An alternative all-sky camera configuration, shown in Fig. 1.37, uses a silvered plastic egg as the primary mirror. The image is reflected up to a flat secondary and then down into the camera lens, mounted under the egg. As with the previous hub cap camera, external support of the optics is eliminated by mounting the secondary on a rod which is coaxial with the camera lens. This places less weight on the rod, which can be shorter, but the effective f number of this system is higher than a conventional hub cap camera.

Since most hub cap cameras are not complete telescopes including mounts, long exposures cause the stars to trail. This has been solved in several cases by placing the entire assembly on a Poncet mount. An alternative approach for small hub caps is to place the assembly on the dust cap of a common telescope which is pointed to the zenith and driven equatorially. This works providing the telescope isn't driven too far from the zenith.

[74] *Sky & Telescope*, June, 1963, p. 340, May, 1965, p. 324, March, 1983, p. 263, and July, 1983, p. 71.
[75] An alternative configuration using a deep dish concave mirror is shown in *Sky & Telescope*, June, 1982, p. 621. A more conventional meteor-patrol camera is shown in the March, 1972 issue, p. 194.
[76] A wide field camera using this principle was shown by Jeff Charles at the 1986 Riverside Telescope Makers Conference. See the *Proceedings* from the conference, p. 79. It was also pictured in *Sky & Telescope*, August, 1986, p. 186. Chris Schur's telescope is described in *Sky & Telescope*, August, 1980, p. 162.

Figure 1.36. Hub cap all-sky camera with single support leg. Photo courtesy of Jeff Charles.

Figure 1.37. All-sky camera made from a silvered plastic egg. Photo courtesy of Jeff Charles.

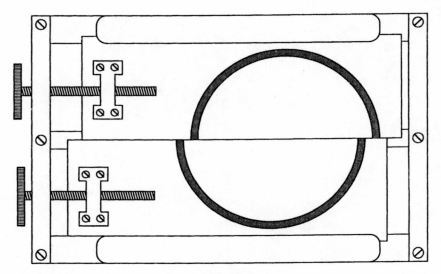

Figure 1.38. Split objective lens.

Miscellaneous optics

The concept of sawing a lens in half may seem abhorrent to most astronomers but several large telescope objectives have been split in order to form an instrument called the Heliometer or Astrometer,[77] as shown in Fig. 1.38. The reason behind the construction of a split objective lens is to be able to move one half of the objective precise distances with respect to the other half in a direction perpendicular to the optical axis. The two images formed will overlap in the eyepiece. This might also be considered a double field telescope. By moving one objective lens with respect to the other, the images will move with respect to each other. If, for instance, a planet is being observed then the lenses could be manipulated until the left edge of the image from the upper lens and the right edge of the image from the lower lens just touch. The angular size of the planet is related to the lens travel required. This arrangement was first suggested by Römer in 1675 and implemented by Dollond in 1754.[78] Although this function has been largely superseded by the bifilar micrometer, the split objective is capable of measuring much wider separations than the micrometer. This has been useful in studies of the apparent diameter of the Sun and Moon.[79]

A second application for a split objective lens is in the visual estimation of star and

[77] *The History of the Telescope*, Henry C. King, Dover Publications, 1955, pp. 153 and 242.

[78] *The History of the Telescope*, Henry C. King, Dover Publications, 1955, p. 150.

[79] A discussion of the split objective can be found in *The History of the Telescope*, Henry C. King, Dover Publications, 1955, pp. 153 and 242. See also *Astronomy*, Fred Hoyle, Crescent Books, 1962, p. 175, and *Sky & Telescope*, March, 1982, p. 302.

Figure 1.39. Twenty foot interferometer spar attached to Hooker 100 inch (2.54 m) telescope at Mount Wilson Observatory. Photo courtesy of the Observatories of the Carnegie Institution of Washington.

nebula brightness. While modern photoelectric techniques have largely supplanted visual observations, considerable effort went into easing the difficult problem of visually estimating small differences in brightness. One way to compare the brightness of a star and a comet or nebula is to defocus the calibration star and compare the surface brightness of the larger image with the comet in question. In order to do this, a split objective was made in which the separate halves could be focused independently. Thus, one part of the objective was focused on the comet. The other was defocused until a star image appeared as bright as the comet. By noting the amount of defocus necessary, an estimate of the brightness difference could be made. The system was also applied to estimating star brightnesses.[80]

One reason for splitting the optics on a telescope is to obtain a larger effective aperture without having the expense of filling in the empty space between the optical elements. A. A. Michelson used this technique, known as interferometry, to determine the disk sizes of several stars. In order to enlarge the aperture of the biggest telescope existing in 1920, Michelson constructed a 6.1 m (20 ft) spar which he attached to the front of the Hooker 100 inch (2.54 m) telescope at Mount Wilson Observatory. The spar, shown in Fig. 1.39, carried plane mirrors which collected

[80] *The History of the Telescope*, Henry C. King, Dover Publications, 1955, p. 296.

twin beams of light separated by up to 6.1 m (20 ft) and directed them into the main optics of the telescope. The rest of the 100 inch aperture was blocked off.

The method of observation depends on the relationship between the resolution of a telescope and its aperture. It also takes advantage of the fact that the star, while normally considered a point source, is really a very small disk. Each point on that disk produces a system of interference fringes at the focus of the telescope. If the star is large then the various sets of interference fringes will overlap and cancel each other out. The effect looks like the disappearance of the fringe pattern. When each star was observed, the fringes were first noted and then the mirrors on the spar were moved farther and farther apart until the pattern disappeared. The separation of the mirrors at that point indicated the aperture of a telescope which could just barely resolve the star's disk. By knowing the separation, the angular size of the star could be determined.

In terms of resolution, the telescope posessed an aperture of 6.1 m but in terms of sensitivity the aperture was only the area of the smaller plane mirrors. Only very bright stars could be measured but the diameters of several stars were determined.[81] Typically, the instrument could measure diameters of a few hundredths of an arc second. The nature of the interference fringes could also indicate if the star were a binary, although for the bright stars tested, previous spectroscopic studies had revealed which stars were doubles so no new double stars were discovered. The relatively small apertures of the plane mirrors and the 6.1 m maximum separation limited the number of stars which could be examined but the experiment proved the technique was valid and worth pursuing.

Following these successes, Michelson and Pease designed an interferometer with a 15 m (50 ft) span. It was mounted as a stand-alone instrument at Mount Wilson Observatory. The telescope looked like a polar telescope with wide wings which supported the moveable interferometer mirrors, as shown in Fig. 1.40. The entire assembly was on an equatorial mount but it could be driven only an hour or so on either side of the meridian. It is a requirement in interferometry that the two small plane mirrors and the auxiliary folding mirrors remain rigid with respect to each other. Unfortunately, the mounting on the huge interferometer tended to sag when pointed to various positions. Mixed results were obtained for several stars and while publishable data were obtained with the instrument, it has fallen into disuse. The remains of the fifty foot interferometer are still at Mount Wilson. The technique of measuring stellar diameters was pursued with slightly different methods, as seen in the interferometer shown in Fig. 4.13.

Starlight from both sources was directed through a polarizer, each beam being polarized perpendicular to the other. The beams were then sent through a second polarizer, acting as an analyzer which was rotated until both stars appeared to be the same brightness.[82]

[81] See *Splendour of the Heavens*, Rev. T.E.R. Phillips and Dr W.H. Steavenson, Robert M. McBride & Co., 1925, Vol. I, p. 69. See also *The History of the Telescope*, Henry C. King, Dover Publications, 1955, p. 337.

Figure 1.40. Michelson's fifty foot interferometer at Mount Wilson Observatory. Photo courtesy of the Observatories of the Carnegie Institution of Washington.

The dialytic or dialyte telescope is a modification of the usual crown/flint refractor combination commonly used to correct for color problems. The basic design was first proposed by Rogers in 1828.[83] At that time, large flint glass blanks were very expensive. One solution was to use a full-aperture crown glass objective and then place the flint glass corrector about half-way down the tube. Since the diameter of the light beam was smaller at this point, a full-aperture flint glass lens wasn't needed. The corrector couldn't be placed very near the focus since it would then require a strong curvature to correct the color aberrations in the shorter distance to the focal plane. Such strong curvatures tend to introduce other optical aberrations so a trade-off was made between the expense of the glass and the complexity of the optical design. The word dialyte means "loose", "separated" or "parted". This type of optical system lapsed in the mid-Victorian era when the price of flint glass dropped but it has been revived by John Wall of Dartford, Kent, England. In his adaptation, seen in Fig. 1.41, he uses three small corrector lenses (two of which are flint glass) situated about half-way down the tube. In his incarnation of the dialytic telescope, he has also shortened the overall optical path by folding it with a flat.[84]

[82] *The History of the Telescope*, Henry C. King, Dover Publications, 1955, p. 296.

[83] *The History of the Telescope*, Henry C. King, Dover Publications, 1955, p. 191.

[84] The instrument is described in *Sky & Telescope*, May, 1975, p. 327.

Figure 1.41. Dialytic telescope. Illustration courtesy of John Wall. The telescope of interest in the figure is the smaller aperture guide scope rather than the main reflector.

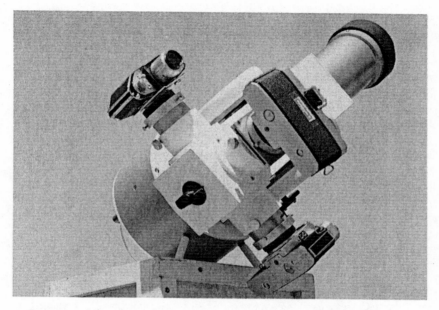

Figure 1.42. Solar eclipse telescope with switchable output beam. Photo courtesy of
Ernie Piini.

Moving the optics around

The concept of moving some optical pieces around with respect to the others isn't
too strange. After all, that's the principle behind most telescope focusing mechanisms. There are times, however, when major components in the optics need to
move for reasons other than focusing. The most common application is in output
beam switching.

Ernie Piini has been chasing eclipses for years. He has learned that during total
solar eclipses, time moves rapidly. If the observer wants to change lenses or film, he'd
better have them already mounted on a separate telescope which is pre-focused and
pointed at the Sun. Otherwise he will waste precious seconds fiddling with
equipment. Rather than pay the freight charges to haul several telescopes to an
eclipse site, Ernie has modified his 8.9 cm (3.5 in) refractor so that the output beam
may be directed to any one of three cameras via a rotating flat mirror controlled by a
knob at the end of the telescope, as shown in Fig 1.42. Each of the cameras has a
different film and filter combination. The telescope might also be classified as a bent
refractor. This scope is also interesting in that the sturdy packing case is used as a pier
for the telescope, thus saving the cost of shipping heavy telescope components.
Ernie has carted the telescope literally all around the world chasing eclipses. He has
designed the telescope to be rugged for portable field use, including "sand proofing"

the mechanical components. It has also been tested by baking the entire assembly at 66 °C (150 °F) in an oven to simulate harsh observing conditions.[85]

There are other occasions where it is desirable to have some of the optical components move with respect to the rest. A case in point is the wobbling secondary mirror. Early infrared detectors were basically photometers in that they measured the brightness of only a single point and did not produce an image. In order to determine the brightness of a star the detector first measured the brightness of the star and then the brightness of a nearby patch of blank sky. The two readings were then subtracted to find the brightness of the star without the intervening sky noise. In order to accomplish this the secondary mirror was rapidly oscillated back and forth so that the detector saw first the sky and then the star plus sky. Typical oscillation periods ranged from 0.1 second to about 0.01 second. Many telescopes such as the Kuiper Airborne Observatory and the Mount Lemmon 1.52 m (60 in) were fitted with wobbling secondaries but the practice is currently being dropped as imaging infrared detectors made of many point source detectors become available.

Convertible telescopes

Convertible telescopes are those whose optical configuration can be changed easily, say from a Cassegrain to a Newtonian. The obvious advantage is owning two telescopes for the price of one. The disadvantages, however, are often cumbersome flip mirrors, the added weight of a second eyepiece holder which can't be used while the other one is in operation and the possibility of compromising optical performance for the sake of convenience. In other words, the primary cannot be optimized to a single secondary or baffling scheme if it is to be used in more than one configuration. Nevertheless, many intrepid designers have undertaken the task. Jim Stevens of the Saguaro Astronomy Club, Metro Phoenix, Arizona, converted a 20.3 cm (8 in) telescope to a Newtonian/Cassegrain configuration as shown in Fig. 1.43.[86] The Newtonian eyepiece holder is attached to the Newtonian secondary diagonal mirror. When using the system in a Cassegrain mode, the Newtonian eyepiece holder rotates so that the eyepiece points toward the object being viewed. This motion rotates the Newtonian secondary out of the optical path, uncovering a Cassegrain secondary which is permanently mounted on a conventional four-arm spider. Jim has added a cover for the Newtonian secondary which is not shown in the drawing.

The 63.5 cm (25 in) space satellite-tracking telescope shown in Fig. 1.15 was capable of conversion to an f/18 Cassegrain. A small Cassegrain secondary could be installed in front of the Ross corrector lens near the prime focus. Although the telescope was designed for conversion and the primary mirror had the required

[85] Details of the mount are given in an article in *Sky & Telescope*, September, 1973, p. 187.
[86] *Proceedings of the Riverside Telescope Makers Conference*, May 26–29, 1989, p. 30. See also *Telescope Making*, No. 37, Summer, 1989, p. 41.

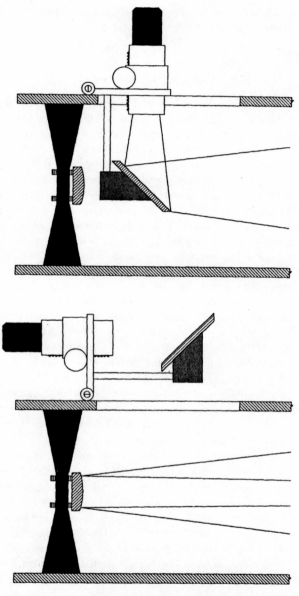

Figure 1.43. Convertible telescope by Jim Stevens.

Cassegrain hole, it was never used in that configuration after the initial acceptance tests. Conversion from prime focus to Cassegrain took less than a minute to accomplish.

Pinholes and optics that don't look like optics

A pinhole with a projection screen behind it is perhaps the simplest optical system possible. Although the image is typically faint due to the high f number, the principle of pinhole projection was known for centuries before the formal introduction of optics. Even today pinhole optics are used for wavelengths where reflective or refractive optics are difficult, such as in imaging X-ray cameras. One modern use of the pinhole telescope is in watching partial solar eclipses where the image is bright. School children often make a box with a white paper inside and a pinhole opposite the paper. When aimed at the Sun, a small image forms on the paper. The box also serves to make a shadowed area so the image contrast is sharper. This is much safer than watching the eclipse directly through fogged film or other chancy filters.

The Kuiper Airborne Observatory crew members once used the pinhole in a snack cracker to project a solar eclipse image and determine the progress of the eclipse.[87] A similar projection, using only human hands, is shown in Fig. 9.4.

While a pinhole may not be considered optics, at least a simple ray-trace diagram showing object and image would be understandable to most people who have worked a little with lenses and mirrors. These work on the principles of refraction and reflection, respectively. There is another optical phenomenon which is capable of producing a magnified image and that is diffraction. Diffraction is produced whenever light passes an edge, as in light passing the vanes of a secondary mirror holder. Some portion of the light is bent toward the edge. Thus, that small portion of light appears to come from a slightly different direction and this makes the resulting image fuzzy. Indeed, the resolution limit of a telescope is determined by light diffracting at the edge of the primary mirror or lens.

The familiar concentric circles of a diffraction pattern in a telescope give a clue as to how to make diffraction work for you instead of against you. The reason for the alternating light and dark rings is due to the wave nature of light. Where waves sum, they appear twice as bright. Where waves cancel each other a dark ring results. Suppose you could identify which portion of the aperture caused the light and dark rings and block off those portions which diffracted light into the bright rings. While in theory this is possible, the trick would work for only one wavelength of light since diffraction is dependent on wavelength. Now turn the problem around and let through only those portions of the waves which add. The image will become brighter. It will appear defocused but the total amount of light will be twice as much.

The proper aperture mask for producing additive diffraction interference is called

---

[87] The incident is described in *Sky & Telescope*, December, 1980, p. 523.

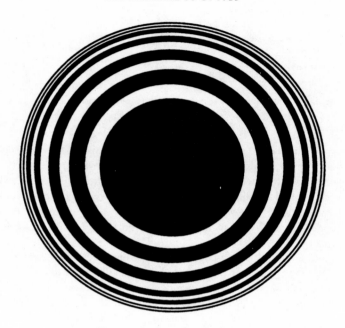

Figure 1.44. Fresnel zone plate.

a Fresnel zone plate, as shown in Fig. 1.44. This should not be confused with a Fresnel lens. Before you snip the illustration out of the book and slap it over the end of your telescope, you should know that this plate is designed to work in the infrared region with a focal length of tens of meters. At visible wavelengths, the lines in the plate are much too fine for reproduction here.

There are wavelength regions, however, where conventional manufacturing methods are quite capable of making a Fresnel zone plate. In the radio region where wavelengths may be millimeters or centimeters long, such diffractive optics are often used. Years ago I happened upon a workman at the National Radio Astronomy Observatory Very Large Array who was sawing a slab of aluminum. The disk, as large as a pizza pan, had fine lines marked on it. Only then did I realize that he was making a radio-frequency optical part with a hacksaw and a coarse file.

Double field telescopes

Double field telescopes are those in which two fields of view are superimposed and projected into a single eyepiece. Typically, both the finder field of view and the image from the main telescope optics are superimposed. While this may seem to present a confusing picture to the observer, a little practice with the system usually results in satisfactory operation. The observer must learn to differentiate the images

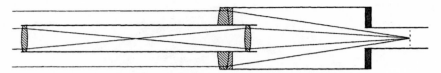

Figure 1.45. Double field refractor telescope. Both wide and narrow field of view images are brought to a common focus for a single eyepiece.

as he views them. This is not difficult since the star images from the finder are generally more crisp. In addition, as the telescope moves with respect to the stars, it becomes obvious as to whether a star is moving rapidly (from the main telescope) or slowly (from the finder). The advantage in this system is that the observer doesn't need to move his head from one eyepiece to another – and on equatorials it seems that the finder scope always winds up on the bottom in some neck-wrenching position.

I have used double field telescopes while looking for Messier objects by star-hopping.[88] For faint galaxies which cannot be seen in the finder telescope, a double field telescope is a real plus. One disadvantage of this system is that the background sky brightness in the eyepiece is the sum of that from the finder plus that from the main optics. This is usually solved by putting a removable opaque cover over the finder to block its light once the object of interest has been found.

A double field refractor telescope, shown in Fig. 1.45, was built by Fred L. Johns.[89] The finder telescope uses the larger aperture lens which has a shorter focal length. The "main" lens is the smaller aperture at the left. The main telescope's image is inverted by the field lens just to the right of the finder lens. Thus, differentiating between the finder and main image is easier because the stars move in opposite directions when the telescope is moved. One problem with this system is that the aperture for the finder telescope is larger than the aperture for the main telescope and thus fainter stars can be seen in the finder than in the main. This telescope was designed, however, to be a combination finder and guide telescope for an even larger telescope used for astrophotography. Thus, the main telescope of this pair is used only for guiding on bright stars during exposures.

A commercially manufactured double field telescope was produced in the mid 1960s by Celestron International. It used a 7.62 cm (3 in) aperture off-axis section from a 25.4 cm (10 in) Schmidt-Cassegrain as the main telescope and a 5 cm (2 in)

[88] In star-hopping, setting circles are not used. The observer starts at some bright star near the object of interest and, using finder charts, slews the telescope past fainter guidepost stars to the object of interest. Often the destination object cannot be seen in the finder so the observer aims the crosshairs of the finder "a third of the way between the faint star and the double" and then goes to the main telescope eyepiece to see if he is at the correct position. This may take several tries.

[89] A complete description of the instrument including the method of coring the finder telescope's refractor lens is in *Sky & Telescope*, July, 1971, p. 42.

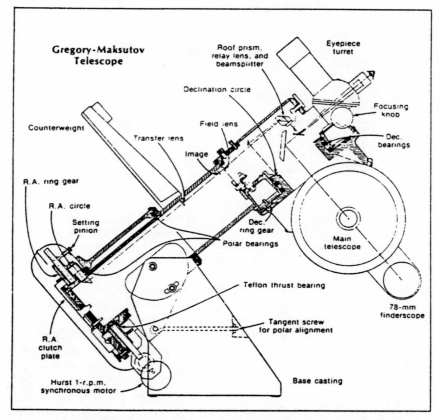

Figure 1.46. Gregory–Maksutov telescope with double fields of view from the finder and main telescopes superimposed. Note that the RA and declination positions are also projected into the eyepiece. Illustration courtesy of John Gregory.

achromat as the finder. A shutter was provided to block the finder field of view once an object had been located. The telescope was not a commercial success and very few were built.[90] Perhaps the public's perception of the problems of differentiating the two fields of view contributed to poor acceptance.

John Gregory designed and built a 21 cm (8.2 in) aperture f/16 double field Maksutov telescope. It differs from conventional Maksutovs in that the light from the secondary is bent by a tertiary mirror to travel up the declination axle to a fourth mirror which directs it up the right ascension axle to the fixed eyepiece as shown in Fig. 1.46. Thus when the telescope moves, the eyepiece stays in a fixed position,

---

[90] Private correspondence, Tom Johnson, founder of Celestron. The telescope appears in an advertisement in *Sky & Telescope*, April, 1965, p. 249.

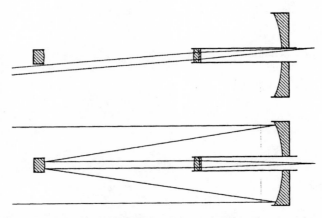

Figure 1.47. Lindsay's double field reflector/refractor telescope. The upper diagram shows operation as a refractor finder telescope. Note that off-axis rays striking the primary will be directed by the secondary to a position outside of the aperture of the field lens. The lower diagram shows rays from the primary or main telescope, used as a Mersenne telescope.

much like a Springfield mount. This ingenious telescope also superimposes the light from a 7.8 cm (3 in) f/8.5 achromat used as the finder onto the image of the primary. Light from the finder is bent by a diagonal mirror toward the tertiary mirror of the main telescope. The beam diameter of the finder telescope is greater than the tertiary mirror size so the finder beam surrounds the main telescope beam. Thus, the main telescope tertiary mirror blocks some of the light from the finder. As an added feature, both the right ascension and declination setting circles are illuminated and images of them can be projected into the eyepiece using a beamsplitter. The beamsplitter can be moved out of the eyepiece's field of view when the setting circle images are not being used.[91]

The previous examples of double field telescopes all had the entire field of view of both optical systems superimposed. A different approach was taken by Wesley N. Lindsay. His refractor/reflector double field instrument is based on a 20.32 cm (8 in) aperture Mersenne telescope.[92] The finder telescope is composed of the eyepiece and a 3.8 cm (1.5 in) aperture f/8 achromat field lens located in the baffle tube as shown in Fig. 1.47. It looks around the edge of the secondary mirror. Thus, its image is an annular field of view (it cannot see a portion of sky at the center of the field). The

[91] A description of the instrument is given in an article by John Gregory in *Sky & Telescope*, February, 1981, p. 165. Several of these telescopes have been built and sold commercially, see *Sky & Telescope*, May, 1982, p. 531. John also built a derivative of the telescope using a 20.32 cm (8 in) aperture f/15 refractor lens. In order to make the tube more manageable, he folded the optical path once. The instrument is shown in *Telescope Making*, No. 32, Spring, 1988, p. 38.

[92] The Mersenne telescope is composed of a paraboloidal concave primary and a paraboloidal convex secondary. The image is not brought to a focus, but leaves the secondary as a parallel bundle of rays. The observer may then look through the perforated primary and see a magnified image at infinity. The design is described in *The History of the Telescope*, Henry C. King, Dover Publications, 1955, p. 48.

main telescope is composed of the Mersenne primary and secondary mirrors which project a parallel bundle of rays toward the field lens in the baffle tube. The image from the main optics fills the entire eyepiece. This might be considered to be a Mersenne telescope with very poor baffling, in which copious quantities of off-axis stray light can enter the field of view. The stray light in this system, however, is in focus and constitutes the finer telescope field of view.

The advantage in this system over the design by Fred L. Johns is that the aperture for the main telescope (the one with the higher magnification) is larger than the aperture for the finder, as in most conventional telescopes. Thus, fainter stars can be seen in the main telescope than in the finder. One difficulty in this design is that it is impossible to block out light completely from the finder telescope without blocking light from the main optics. This means that the background brightness is increased as seen in the eyepiece.[93]

Lens/mirror combinations

In fast, wide field telescopes the differences between spherical and aspheric primaries become greater as the f number decreases. The costs of grinding an accurate low f number parabola become greater than the costs of introducing extra optical elements such as field flatteners, coma correctors, etc. It is in this regime that lens/mirror combinations are most often found. Combinations of mirrors and lenses are often used to avoid the problems of grinding aspheric optics. In such cases, a strong spherical primary mirror is combined with weak refracting optics which correct for aberrations. The commonest lens/mirror combinations are the Schmidt telescope and the Maksutov telescope. Both of these designs have been employed widely in both the professional and amateur communities. They have also been produced commercially by the thousands. There are, however, several other alternatives possible.

One short cut in mirror making is to undercorrect the primary, making it ellipsoidal rather than parabolic, and then use a spherical secondary. Horace Dall perfected the design, known as a Dall–Kirkham telescope in the 1930s.[94] Dall took the design further and added an erecting relay lens between the secondary and the eyepiece. The erecting lens was located part-way up the baffle tube. This served the additional purpose of keeping stray light out of the field of view. A typical Dall–Kirkham telescope actually has no need for a closed tube or baffles.[95] Dall took advantage of this when he made a folding telescope for travellers which has no tube and only a minimal secondary support. The design has been copied often.[96]

---

[93] There are several approaches to decreasing the background light from the finder, as discussed in Lindsay's article in *Sky & Telescope*, February, 1965, p. 112.

[94] *Sky & Telescope*, January, 1962, p. 48. See also *Sky & Telescope*, October, 1986, p. 410.

[95] An excellent example of a Dall–Kirkham telescope with an open trusswork tube is at the Dayton Museum of Natural History, as shown in *Telescope Making*, No. 3, Spring, 1979, p. 26.

[96] See *Sky & Telescope*, March, 1968, p. 182.

The addition of a relay lens in the central baffle tube of a Cassegrain serves at least two purposes. First, it can be used to image the aperture defined by the secondary mirror onto baffles behind the relay lens. This means that stray light entering the telescope near the secondary mirror but so far off axis as to miss the primary mirror cannot find its way down the baffle tube to the image plane. The relay lens will deflect it to the side of the baffle tube before reaching the image plane. As with the Dall–Kirkham telescope, little or no outer telescope tube is required to keep out stray light. The central baffle tube also holds the lenses in place. The second purpose for the relay lens is to correct those aberrations which are left over after the primary and secondary mirrors. This may be done empirically, as has been done in the past, or by the application of computer-driven ray-trace programs.[97] One advantage of applying the final corrections to the relay lenses is that they are typically small glass surfaces and are therefore much easier to generate than the larger aspheric surfaces of the primary and secondary mirrors.

There are a variety of designs which use relay lenses, in much the same way that a periscope operates, in order to bring the image out of a basically classical Cassegrain telescope. Most of them erect the image or turn it right side up as in Dall's design. Many also incorporate at least rudimentary aberration correction within the lens system, such as the Dilworth telescopes. Typically, these telescopes are character-ized by an extended tube between the primary mirror cell and the eyepiece position.

In the never-ending quest for more aperture, wider fields of view and better resolution, several factors are fighting each other. Larger mirrors are harder to figure to a parabola than smaller ones. Thus, resolution suffers or the cost of figuring increases. As wider fields of view are used, especially for photography, the off-axis aberrations increase markedly. One solution to this is to use a spherical mirror and correct the aberrations elsewhere. As with the catadioptric Cassegrain types, a small field lens or corrector may be inserted in the optical train in front of the eyepiece. In Newtonian systems, the logical place is just in front of the secondary mirror as shown in Fig. 1.48.[98] The spherical aberration of the corrector can be made opposite to that of the mirror and equal in magnitude. Although there are a myriad of ways to correct for the aberrations in simple lenses and mirrors, one of the more ingenious is the Schupmann or medial telescope. Originally conceived by Ludwig Schupmann, the design uses a weak lens and a Mangin mirror to correct for spherical aberration, coma and other problems in a refractor.[99] Although the type is produced in several

[97] One of the earlier computer optimized relay designs was described by Richard Buchroeder in Sky & Telescope, April, 1968, p. 249.

[98] Robert T. Jones described an all-spherical Newtonian in which two corrector lenses were placed just in front of the secondary, see Sky & Telescope, September, 1957, p. 548. Sylvain M. Heumann took the design and scaled it up to an all-spherical 25.4 cm (10 in) Newtonian with a negative correcting lens (air spaced doublet) in front of the secondary, as described in Sky & Telescope, October, 1962, p. 231. A commentary by Berlyn Brixner to Jones' design appears in Sky & Telescope, August, 1966, p. 103.

[99] The basic mechanism of the Schupmann telescope is shown in an article in Sky & Telescope, September, 1956, p. 512. Schupmanns are also discussed in the issues of August, 1956, p. 434 , October, 1964, p. 204, March, 1983, p. 275, November, 1983, p. 446, November, 1984, p. 405 and January, 1986, p. 97.

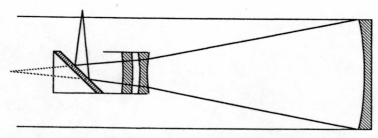

Figure 1.48. All-spherical catadioptric Newtonian telescope. The corrector lenses just in front of the secondary mirror correct for aberrations in the primary.

variations, a typical Schupmann uses a folding flat mirror near the eyepiece to direct the light through a lens with spherical surfaces. Both positive and negative lenses have been used depending on the residual aberrations in the objective lens. The beam then encounters a Mangin mirror which is silvered on the back side. The light thus passes through the glass of the mirror twice on its way to the eyepiece. In some designs the light is bent once more to an eyepiece while in others the Mangin mirror is tilted slightly to send the light directly to the eyepiece.

The largest Schupmann is the 40.6 cm (16 in) solar telescope at Sacramento Peak Observatory. Although the Schupmann concept is usually applied to refractors with a single-element objective lens, it has been used for doublet objectives. The advantage in this approach is that all of the complex optics required for correcting aberrations are near the eyepiece end of the telescope and thus they are smaller and less costly. One disadvantage, seen in Fig. 1.49, is a delicate optical assembly which sticks out at right angles from the tube. With a judicious selection of lens and Mangin mirror curvatures, the design can correct for residual color problems very well. Thus, Schupmann refractors are often used for planetary and solar observations.

Lest the reader think that optical design is "finalized" he is encouraged to examine A. W. Wilkinson's article in the *Journal of the British Astronomical Association*, Volume 88, No. 1, December, 1975 p. 35 in which several new lens/mirror configurations are suggested. With the introduction of inexpensive computers and ray-tracing algorithms, it is expected that new optical designs will surface continuously.

One driver of new optical designs might be the advent of computer-controlled optical machining. This emerging technology uses diamond cutters to shape mirrors and lenses. The current state of the art is capable of making diffraction-limited optics in the infrared region. As tolerances improve, it is expected that visible light diffraction-limited mirrors and lenses will be feasible. Since the computer controlling the diamond lathe uses a mathematical equation to guide the tool, it is just as easy to specify an aspheric surface as a spherical one. In the past, opticians tried to design using all-spherical surfaces and occasionally, if there were no other alternative, they might add an aspheric to meet specifications. Now opticians may use aspherics from the start of the design. This will certainly influence the way telescope makers approach optical problems.

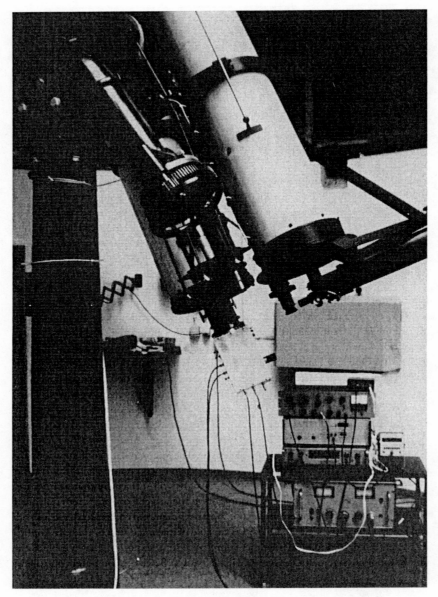

Figure 1.49. Schupmann (medial) refractor. Illustration courtesy of C. Wolter.

Occasionally there will be a letter in the popular journals that innovative optical design is now finished. Just remember that in the 1840s there was a move to close the Patent Office because, at the time, it was felt that everything which could possibly be invented had already been thought of.

Eliminating diffraction spikes

In many astrophotos there are cross-shaped extensions to the brighter stellar images. While some people see these artifacts as pretty and esthetically pleasing, they are the bane of conscientious astrophotographers. The cause of diffraction spikes is light interacting with the surfaces of the secondary mirror supports. Typically, there are four spider supports in most Newtonians and Cassegrains so there are four spikes. The physics of the diffraction phenomenon, however, is such that light interacts with any edge it passes near. Assuming that the four vanes of the spider are in the North, South, East and West directions inside the telescope tube, it is the North vane and the South vane that cause the spikes in the East–West direction. The Eastern edge of the North vane causes half of the Western spike and the Eastern edge of the Southern vane causes the balance of the Western spike. Similarly, the Western edges of the Northern and Southern vanes cause the Eastern spike. Thus, if only two supports (Northern and Southern vanes) were used, as is done in some telescopes, there would still be spike images in both the East and West directions. In this case there would be no Northern and Southern spikes in the image.

In an attempt to eliminate diffraction spikes, some telescope designers have incorporated curved secondary mirror holders.[100] An excellent example, shown in Fig. 1.32, is Ed Danilovicz's 31.1 cm (12.25 in) f/6.4 bent Newtonian.[101]

The diffraction phenomena are still present in the telescope but because of the curved edges, the light from the diffraction spikes is smeared out around the star in all directions and has become less noticeable. If you measured precisely the star image and the amount of light immediately adjacent to the star image, you would find that the diffraction is still present. It should be noted that light also diffracts off the edge of the aperture and this is what causes the "diffraction limited" image.

At least one telescope designer has reasoned that if he eliminated one of the spider vanes, he would eliminate one of the spikes in the image. The 40.6 cm (16 in) aperture telescope shown in Fig. 1.50 is one such. Remember, however, that the diffraction effect is caused by the edges of the vanes and with three vanes placed 120° apart, there are six edges each 60° apart. Astrophotos taken with this telescope show six crisp spikes placed 60° apart.[102]

---

[100] A one-armed curved secondary holder is shown in *Sky & Telescope*, September, 1989, p. 313. Care must be taken in such designs lest the secondary holder become a tuning fork, vibrating all night.

[101] Other curved secondary mirror holders can be found in *Telescope Making*, No. 33, Summer, 1988, p. 37. See also *Sky & Telescope*, February, 1969, p. 113, December, 1975 p. 427, April, 1976, p. 278, June, 1979, p. 584, and May, 1982, p. 523.

[102] For a discussion of diffraction patterns seen with various secondary support systems, see *Telescope Making*, No. 7, Spring, 1980, p. 4.

Figure 1.50. Secondary mirror holder with three vanes. This produces six spikes of diffraction.

I have seen at least one case of diffraction spikes being introduced purposely. One owner of a Celestron C-8 Schmidt-Cassegrain wanted an astrophoto to be used on a Christmas card. Since the C-8 uses the Schmidt corrector plate to support the secondary mirror, there are no spider vanes. In order to produce the cross-shaped effect on the card, he stretched two threads across the aperture at 90° angles. The resulting photos show nice crosses.

2

Telescope mounts

The basic idea of a reflecting telescope is to hold about 0.01 grams of aluminum (the Palomar 5 m (200 in) Telescope uses 5 whole grams) in a very precise place with respect to an eyepiece. The rest of the telescope is just support mechanism and accessories. While some mounts may be massive marvels of mechanical engineering, the prime purpose of the mount is to keep the telescope tube off the ground and help you to point it accurately. If your needs are for very wide field casual stargazing, then the best mount is your own two hands holding a pair of binoculars. If, on the other hand, you have precise pointing and tracking requirements then the mount may easily cost as much as the optics. Indeed, for some high speed tracking mounts, the optics are a very small part of the expense. Unfortunately, in many telescopes so much emphasis is paid to the optical quality that the mount is considered only as an afterthought. I have seen many fine refractors which are attached to such flimsy wooden tripods that they are virtually unuseable. There have also been a few well-made mounts – they would win awards as beautiful hand-crafted furniture – which are attached to such poor optics that they are also unuseable. I have seen one computer-driven telescope which accurately and rapidly slews to any point in the sky. The software is superb and useable by those who are intimidated by computers. If the optics had been even adequate it might have been a good telescope.

As with any engineering effort, trade-offs are made in mount design and a balance is struck between form, fit, function, cost and even esthetics. There is certainly something appealing about a design which functions with a minimum of materials as seen in the bare bones telescope shown in Fig. 1.16. From a design standpoint efficient use of materials is always a plus. In the case of a telescope, fewer parts usually means lower weight and increased portability. In this chapter we will see many methods of supporting telescope optics. The sections are arranged by the number of axes incorporated. While most telescopes have two axes, there have been telescopes made with from zero to four moving axes.

Zero-axis telescopes

While most telescopes are built to scan all or most of the available sky, there are four cases where no motion is required by the telescope. The first is in the operation of a zenith telescope, as shown in Fig. 1.7. These are usually used in the precise determination of either time or position of the observatory. Some have been used to measure either the motion of the observatory due to continental drift or the wobble of the Earth's axis. A select band of stars which passes directly over the observatory is all the telescope ever observes.

A second reason for staring at one point in the sky is to monitor Polaris for atmospheric seeing measurements. The reason for this is that such a seeing monitor essentially needs no tracking mount and can make automated, unattended observations. A typical installation would include a refractor mounted firmly and pointed at the pole. A beamsplitter near the image plane would send half of the light to a photometer which measures the brightness of Polaris. Since the pole star is about 50 arc minutes from the true center of rotation, the field of view of the photometer must be wide enough to see the star at any place in its daily march around the pole. The other half of the light would be masked by a reticle at the image plane. The reticle is opaque except for a thin ring corresponding to the expected path of the pole star. The width of the ring is so small (barely a clear line etched on the opaque material) that its width corresponds to only about one arc second. If the seeing is worse than this limit then much of the light from Polaris will fall on the opaque material and a photometer behind the reticle will sense a drop in the light. If the light from both photometers decreases then clouds have blocked the view. Thus, both seeing and cloud cover can be measured automatically.

In some systems the opaque material of the reticle is a reflective metal coating. Light from the center of this reflective coating can be directed into a third photometer which measures the background sky brightness of the night sky. Most major observatories install such seeing monitors at candidate telescope sites before constructing expensive telescopes. It prevents finding out that your ten million dollar observatory building will have to be moved to the next ridge if you hope to see anything clearly.

The third reason is for very deep penetration surveys of galaxies. The 1.8 m (70.9 in) CTI (CCD Transit Instrument) telescope at Kitt Peak stares at the zenith and records thousands of faint galaxies each night in a narrow band of sky which passes overhead. The telescope uses an integrating charge-coupled device (CCD) which records stars to Mv 20. By comparing a galaxy's image with the same image seen the previous night, supernovas can be detected. In addition, the telescope is looking for quasars. While the telescope is nearly a zenith instrument, it was tilted slightly and views near the zenith on the meridian. The reason for tilting was to include a swath of the Coma cluster of galaxies. Each night, the telescope makes a single picture which is 8.25 arc minutes wide and 8 hours in right ascension long.

The fourth case is a calibration horn antenna used by the National Radio Astronomy Observatory (NRAO) at Green Bank, West Virginia, as shown in Fig. 2.1. While the horn is normally used for calibrating the sensitive receivers used at NRAO, it has also been used to observe the powerful radio source Cassiopeia A

Figure 2.1. Calibration horn telescope (horn antenna) at NRAO. The 36.6 m (120 ft) long tube tapers from a 21 m² mouth to 0.02 m² at the receiver. Photograph courtesy of the National Radio Astronomy Observatory, Green Bank, West Virginia. .

which sweeps past the horn once a day.[103] Such a radio telescope is analogous in the visible light regime to using a long tapered shiny tube for an optical system. The observer simply measures the total radio energy at the wavelength of interest which spills out the small end. The resolution of the telescope is, at best, only about 6°. In order to form an image with this type of telescope, it is pointed to various areas in the sky and the radio flux at each position is recorded. Each data point is then plotted or graphed to form an impression of the strength of radio sources at various locations in the sky. Since this particular telescope is fixed to the Earth, it sweeps out a circle at the declination of Cassiopeia A and can't look at any other sources unless it is dismantled and re-erected pointing in some other direction. While this may seem crude, it is effective and the large aperture, relative to optical telescopes, is much less expensive per square meter. In reality, there are two fixed telescopes in Fig. 2.1. On top of the

[103] *Sky & Telescope*, December, 1974, p. 352.

receiver building is a second smaller horn which points at the zenith. This is a calibration antenna which looks at the cold sky for comparison with the signal from the main antenna horn.

One-axis telescopes

One-axis telescopes are usually meridian transit instruments. The single axle is arranged in an East–West direction. The telescope is then free to swing from the South horizon through the zenith to the North horizon. In order to observe any particular star, the observer must wait until it passes through his meridian. Such instruments are often used to precisely determine star positions by noting the time at which a star passes the meridian. Since these telescopes are used to make very accurate measurements of star positions, they have massive foundations and axles in order to minimize the errors due to flexure in the mount. Their setting circles are often large in order to accurately determine the star's declination at the time of transit.

The telescope shown in Fig. 2.2 is used by the US Naval Observatory and although it appears to be an older telescope, it has been modernized as the technology progressed. Observers used to lie on a couch between the piers and call out to an assistant the exact time when a star's image crossed a wire in the eyepiece. In today's operations, the observer's eye has been replaced by a photometer and the assistant no longer records data with pen and paper. A sophisticated electronic data recording system notes events with respect to an atomic clock and logs data automatically.[104]

A curious one-axis telescope is James Bradley's 3.8 m (12.5 ft) focal length zenith sector of 1727. It was originally erected at his home and was later moved to Greenwich Observatory in 1748 after he became the Astronomer Royal in 1742. The telescope operated as a meridian transit but could view stars only within a few degrees of the zenith. The objective end was suspended from a pivot on the roof and the tube hung down into the observing room. Although it had a long focal length, the aperture was fairly small and the eyepiece allowed only high magnifications. Equipped with a finely graduated arc sector at the eyepiece end, the telescope measured zenith distances of stars accurately. The instrument was originally designed to determine the parallax of nearby stars but instead it was the first to possess sufficient precision to detect the aberration of starlight due to the Earth's motion.[105] The instrument may have been made for Bradley by George Graham.[106]

---

[104] The six inch transit instrument is described in an article in *Sky & Telescope*, February, 1966, p. 69. Several innovations have been added to the instrumentation since the article was written.

[105] *Mercury*, the journal of the Astronomical Society of the Pacific, May/Jun, 1989, p. 73.

[106] See *The History of the Telescope*, Henry C. King, Dover Publications, 1955, p. 110 and *Astronomy*, Fred Hoyle, Crescent Books, 1962, p. 150. A picture of the restored instrument is in *Sky & Telescope*, July, 1970, p. 8.

Figure 2.2. Six Inch Transit Instrument. Photo courtesy of US Naval Observatory.

Figure 2.3. A replica of Karl Jansky's antenna. The replica was built by Bell Laboratories and donated to the National Radio Astronomy Observatory in 1966. Photo courtesy of NRAO.

While most one-axis telescopes move only in elevation and act as transit instruments, at least one moves only in azimuth. This is the first radio telescope, designed by Karl Jansky of the Bell Laboratories in 1931.[107] He was assigned to look into radio noise phenomena which interfered with long distance radiotelephone calls. While the radio antenna he built, shown in Fig. 2.3, may not look like a telescope, it is indeed a radio telescope in the truest sense. The telescope itself is made of wood and rotates on automobile tires. It is pointed manually by pushing on the mount. It has an aperture of 30.5 m (100 ft.) by 3.7 m (12 ft.). While it doesn't have a f number or focal length in the conventional sense, it does have a field of view of about 45°. It is most sensitive to radio waves striking it broadside and least sensitive to radio waves passing along its length. With the best radio receivers of the day, Jansky was able to detect emissions from the center of our own galaxy, opening a whole new field of astronomy.[108]

The detection of radio waves from non-Earth sources was a tricky thing in those days. The radio receiver indicated the strength of the radio waves on a meter or a strip chart. When the antenna was pointed at an azimuth where the center of the

[107] See *Sky & Telescope*, December, 1974, p. 362. The telescope is also discussed in *The Amazing Universe*, Herbert Friedman, National Geographic Society, 1975, p. 135.

[108] Karl Jansky is honored today by the naming of the international unit of radio flux density, called the jansky, which is equal to 10^{-26} watts per square meter per hertz.

galaxy would soon rise, the needle position might waver uncertainly and then increase over a few minutes while the radio source rose above the horizon. Only by repeating the experiment daily and noting that the increase occurred about four minutes later each day was Jansky convinced that he was receiving radio waves from space. He labored for quite some time eliminating other sources of radio energy such as man-made emissions and natural radio noise in the upper atmosphere.

After the discovery of radio emissions, Bell Laboratories did not pursue radio astronomy heavily until the 1960s. Karl Jansky went on to other problems in radio engineering. Radio astronomy lay dormant until after World War II when amateur astronomer and professional radio engineer Grote Reber revived it in his back yard.[109]

Reber's original 9.6 m (31.4 ft) aperture f/0.63 radio telescope was mounted as a one-axis meridian transit instrument as shown in Fig. 2.4.[110] He built it by hand from available materials. In addition, he built the radio receivers which were specially designed for astronomical observations. There were no other radio receivers like his in the world at the time. While Jansky may have made the first radio astronomy observations, Reber is generally credited with engineering the techniques of observation to workable practices which other astronomers could use regularly and reliably.

The telescope had a sheet metal mirror shaped into a rough parabola. It was used at a wavelength of 1.85 m to correlate the radio noise discovered by Jansky with the distribution of matter in the Milky Way. The drive system was fully manual and the usual method of observing was to set the telescope at some particular declination and then record the radio intensity observed for several hours on a strip chart while the sky passed in front of the antenna. Later, Reber would try to sort out the real radio signals from those generated by passing cars with faulty ignition cables.[111]

There has, from time to time, been debate on whether astronomical radio instruments are really telescopes at all. Like the calibration horn antenna shown in Fig. 2.1, Jansky's and Reber's devices do detect objects at a distance and thus satisfy the definition of a telescope. They do use a mirror (the ground plane) and they are sensitive in a particular direction so that the source of the detected radio energy can be identified. The angular resolution of Jansky's telescope was crude and its

[109] Early reports of Reber's work appear in *Sky & Telescope*, February, 1945, p. 7 and April, 1949, p. 139. A history of his work is given in the July, 1988 issue, p. 28. See also *The History of the Telescope*, Henry C. King, Dover Publications, 1955, p. 436.

[110] The telescope was later rebuilt and an azimuth axis was installed. Reber's antenna has since been moved to NRAO Headquarters at Greenbank, West Virginia.

[111] As a small boy growing up in rural Illinois a few kilometers from Reber's house, I remember seeing the dish antenna he built in Wheaton. At the time I was told that there was a crazy man over in the DuPage County Seat who said he was listening to the stars on his radio. It wasn't until I met him years later in a chance encounter at the National Radio Astronomy Observatory Very Large Array that I realized who he was and the significance of that antenna.

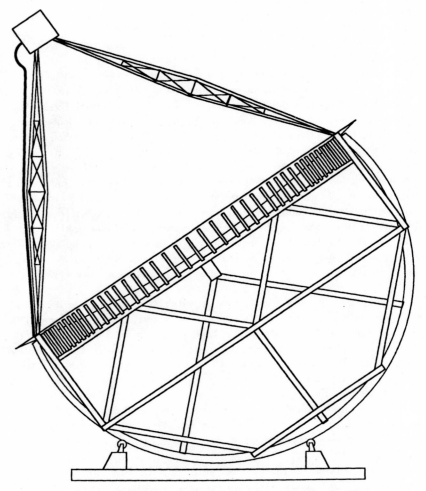

Figure 2.4. Grote Reber's original radio telescope as it appeared in Illinois.

sensitivity was poor by our standards but it was the largest and most sensitive radio telescope in the world when it was the only one. No, it doesn't look like any conventional telescope we're used to but the first Newtonian telescope didn't look like any telescope of its age either.[112]

[112] For an interesting perspective on the introduction of a new type of telescope, Newton's paper, originally published in the *Philosophical Transactions of the Royal Society of London*, has been reproduced in *Telescope Making*, No. 9, Fall, 1980, p. 4.

Two-axis telescopes

Two-axis telescopes are, of course, the commonest type. With two axes one can point the telescope to any position in the sky. The arrangement of the axles can vary considerably, however. Normally the axles are perpendicular to each other. Well, they're supposed to be perpendicular anyway. I do know of one yoke-mounted equatorial where the orthogonality of the axles is off by a full two degrees. Originally, the telescope was pointed manually without benefit of the declination setting circle which would be up to two degrees off. During the 1970s the telescope was fitted with precision encoders and a computer-control system. The computer software now automatically adjusts for the error in the declination axis and points the telescope correctly.

Altitude-azimuth (alt-az) mounts

Alt-az mounts are probably the easiest to construct. The bearings must withstand side or end forces in only one direction at a time. While they require simultaneous nonlinear motion in both axes to track the stars, this is not a problem with computer control of the drive motors. It is also not a problem if the telescope is small and hand guided, as in the popular Dobsonian visual telescopes.

Alt-az mounts were used in the past for the very largest telescopes. Prior to the middle 1800s it was nearly impossible to fabricate bearings and gears larger than about 1.5 m (59 in) in diameter. Telescopes with lengths of 20 m or more, weighing tens of tons simply could not be put on equatorial mountings requiring large precision mechanical parts. Thus, Herschel's forty foot telescope, shown in Fig. 1.13, was built as an alt-az instrument.

A similar problem faced Reverend John Craig in south London. He ordered a doublet refractor lens of 60.96 cm (24 in) aperture and, upon completion in 1852, mounted it as shown in Fig. 2.5.[113] The telescope was supported on a tower with a rotating top and a circular track for the base ring. This system is useable if you just happen to have a water tower in your back yard. The cigar-shaped riveted sheet-iron tube is really a double cone and this makes it stronger than a cylinder of similar dimensions. The tube, nearly 23 m long, was about 1.3 m in diameter at its widest point. It is unclear from drawings and descriptions whether the whole tower or just the tower top rotated in azimuth. The base ring was made of rails so that a carriage could move around the tower to different azimuths. As the telescope was adjusted in

---

[113] Although probably the largest refractor in the world at the time, the lens was of such poor quality that the central part of the lens had to be covered by opaque material according to *The History of the Telescope*, Henry C. King, Dover Publications, 1955, p. 254. The telescope is also described in *Sky & Telescope*, July, 1982, p. 12.

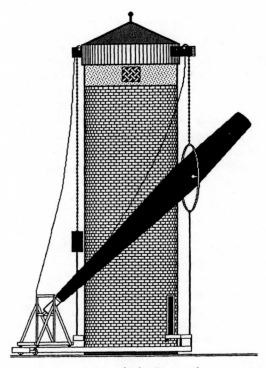

Figure 2.5. Reverend John Craig's telescope.

elevation, a wheeled dolly moved toward or away from the tower on a second set of rails mounted on the carriage. While the telescope appears to have been pointed coarsely using the chain hoist, carriage and dolly, fine tracking was accomplished by moving the eyepiece with respect to a framework which was supported by the dolly.

Owing to poor quality optics the telescope wasn't used much although it did serve as a local landmark for Wandsworth, observable by passengers on the nearby railway line. The exaggerated claims made by the tabloids of the day of wonderful sights never materialized. Within a few years the structure was moved to Leamington Spa. This telescope has often been called "a complete failure" but the instrument may be useful as an example of how not to build a telescope.

One exception to the rule that an alt-az telescope requires motion in both axes to track the stars is if the telescope is located at the North or South Pole. In those cases, an alt-az mount is also an equatorial mount, as shown in Fig. 2.6. This telescope is designed to make photoelectric photometry measurements of stars whose brightness varies over a period of days. During the six months of darkness each year, the telescope can continuously monitor a star. The South Pole has a harsh climate and it is very expensive to support observers there. Thus, the telescope was designed to be

Figure 2.6. South Pole telescope located at the Amundsen–Scott Station. Photo courtesy of MaryJane Taylor, University of Florida.

computer controlled, relaying its observations via satellite to its creators sitting in sunny Florida.[114]

The alt-az mount is generally more compact than the equatorial. Thus, it has a lower moment of rotational inertia and can be slewed rapidly around the sky. While most astronomers do not mind spending a minute or more lining up on each object, there are some applications where the telescope must be pointed quickly and accurately. One is in space satellite tracking systems and another is in the study of transient astronomical events.

Gamma-ray bursters are enigmatic astronomical objects. They were first detected by instruments designed to monitor nuclear explosions. Located on several spacecraft, the detectors found occasional bursts of gamma-rays. A coarse position of the source objects could be made but the precise object couldn't be pinned down.[115] It has been theorized that gamma-ray bursters are related to optical bursters. In order to locate the sources of the bursts which last just a few seconds, two telescopes will be operated together at Kitt Peak. The Massachusetts Institute of Technology Explosive Transient Camera (ETC) consists of several small CCD television cameras with wide fields of view attached to a single equatorial mount. The ETC stares at a large section of the sky, comparing the current scene with past scenes, taken several seconds earlier. A computer will detect optical bursts (after weeding out airplanes, meteors and moths flying near the camera). The system is designed to operate without an observer. The ETC can determine the position of the optical burst to within two arc minutes. This coarse position of an optical burst in progress will then be sent to the Rapidly Moving Telescope (RMT), developed by NASA Goddard Space Flight Center. The RMT, shown in Fig. 2.7, uses a fixed 18 cm aperture telescope which looks down onto an alt-az mounted mirror which can act as a siderostat. The mirror is computer controlled and can be slewed to any position in the sky within one second of time, to an accuracy of 2.5 arc seconds. A CCD TV detector in the RMT then makes a series of exposures in order to record the precise position and light curve of the burster. While the RMT is waiting for the ETC to detect an event, the RMT will stare at one of the suspected positions of an optical burster in the hope that it will flare again. Once the precise positions of these objects are known, larger telescopes with spectrographs can then be trained on the objects in order to determine the nature of the phenomena of optical and gamma-ray bursters.

Equatorial mounts

The equatorial mount is, of course, the most common telescope mounting. The advantages of tracking the stars by driving only one axle are obvious, although

[114] A description of the South Pole Optical Telescope (SPOT) is in *Sky & Telescope*, October, 1988, p. 351.
[115] Some historical sleuthing in archive photographic plates has turned up some likely candidates for several of the gamma ray bursters but none have been observed to flare while being watched by high resolution optical or gamma-ray telescopes.

Figure 2.7. Rapidly Moving Telescope with alt-az mounted mirror. Photo courtesy of Scott Barthelmy, NASA Goddard Space Flight Center.

recently with the advent of computer-driven telescopes, alt-az mounts are becoming more popular. While the origin of the equatorial mount is probably lost in history, there have been many forms of it.[116] The angle that the polar axle makes with the local vertical depends on the observer's latitude. In the extreme cases, an alt-az mount is equatorial if situated at the poles and an el-el mount is equatorial if placed at the equator.

Since the purpose of a mount is to hold the optics and cameras or eyepieces, it doesn't necessarily have to be beautiful. It merely needs to be functional. Some of the more functional mountings have been made for total solar eclipse expeditions. These are often specially made for one observation at one particular latitude and then abandoned. A typical example is shown in Fig. 2.8. This mount, used in Kenya during the 1980 eclipse is really an equatorial mount. The polar axle is nearly horizontal because the path of totality was only a few degrees South of the Equator. The beam holding the cameras could be tipped with respect to the polar axle in order to adjust declination. Note that the polar drive is a lead screw turned by a small motor. While such a drive develops tracking errors near the ends of its travel, a total solar eclipse lasts only minutes so a full worm wheel and worm gear are not needed. The mount carries two still cameras, a movie camera and a finder scope. Note that two of the

---

[116] The mounting of an instrument with respect to the celestial sphere predates the invention of the telescope. A typical example is Regiomontanus' torquetum from about 1470. For a description of this instrument, see *The History of the Telescope*, Henry C. King, Dover Publications, 1955, p. 10.

Figure 2.8. Solar eclipse camera mount used in Kenya.

cameras are remotely controlled. This is because during a total solar eclipse the observer has his hands full operating one camera. Time flies during the moments of totality and many observers prefer to rely on automatically sequenced cameras rather than become flustered and ruin the exposures.

Elevation-elevation (el-el) mounts

In an el-el mount, the usual alt-az axes have been tipped 90° so that the telescope looks like an equatorially mounted telescope located at the equator. One advantage of this design is that the outer bearings do not have to take both end and side thrusts, as the right ascension axle in most telescopes does. As with the alt-az mount, the telescope cannot track the sky with motion of a single axle unless it is located at the equator. While the el-el mount tracks the sky, the field of view rotates in the eyepiece just like an alt-az telescope. With the advent of computer-controlled telescopes and electronic imaging, these inconveniences diminish in importance. One of the reasons for using an el-el mount rather than an alt-az mount is that there is no "dead zone" at the zenith. As a star passes through the zenith of an alt-az mount, the azimuth slew-rate goes to infinity. In practical applications, this means that there is a zone several degrees wide at the zenith where the telescope cannot track a star accurately. The el-el mount also has "dead zones" but they are at the North and South horizons, locations unlikely to be observed often. It should be noted that equatorial mounts

Figure 2.9. El-el mounted FAR telescope by Max Bray.

also have a dead zone at the celestial pole. Observations near the pole are characterized by large motions in right ascension with little resulting movement of the field of view. Since astronomers seldom need to track anything moving through this zone, it is not usually a problem.

A classical el-el telescope was built by Max Bray of Phoenix, Arizona. The instrument was designed for the Funds for Astronomical Research (FAR) foundation as a quarter-scale model for a proposed spectroscopic telescope. The 30.48 cm (12 in) aperture instrument, shown in Fig. 2.9, is computer controlled. It has a moving tertiary mirror which directs the beam out to a large fixed spectrograph at the end of the outer axle. The reason for this arrangement, which is more of a coudé than any other type, is to allow the telescope to move without having to carry the weight of the spectrograph. The problem of field rotation doesn't matter in this system since it is used as a spectrographic and not a photographic instrument. This telescope is also shown in Fig. 1.30.

An el-el mount can also be useful when it is desirable to have the space below the telescope clear of obstructions. Randall Wehler constructed the mount shown in Fig. 2.10 to hold his low power 11.43 cm (4.5 in) f/4.5 refractor telescope, made from a surplus aerial camera lens.[117] The mount is adjustable in height by moving the outer

---

[117] For a complete description of the mount including construction details, see *Sky & Telescope*, January, 1986, p. 95.

Figure 2.10. Wehler's el-el mount with a wide field refractor mounted. The cross-arm may be moved up and down to accommodate comfortable viewing from the lounge chair. Photograph courtesy of Randall Wehler.

elevation axles to different hooks on the posts. In order to make large changes in elevation, the back of the lounge chair is tilted or the entire chair is scooted back and forth. He is considering putting wheels on the lounge chair. No drive motors are required since the 9° field of view allows the observer to track objects easily by hand. The entire mount is made of wood and pipe fittings. It is balanced so that it will remain in any position when the observer isn't touching the scope but the friction on the bearings isn't so great as to make movement difficult. Randall uses the telescope for visual rich-field (wide-field, large aperture) sweeping and comet hunting. This is not a scientific instrument, but rather one born in the interests of laziness and comfort. Those requirements often make for good design.

Three-axis telescopes

You may ask why add an extra axis? Most portable German mount telescopes actually have a third axis for setting the elevation of the polar axis but, once set, the third axis is clamped and forgotten. There is, however, another more practical reason for one class of observing problem. A three-axis mount can track any great circle in the sky with motion on a single axis. Why do that? To track Earth satellites, of course (doesn't everybody?). While Earth satellites do not exactly describe great circles, they approximate the motion. Since a satellite can pass through the zenith where an alt-az telescope would have to develop an infinite slew-rate in azimuth, those mounts are not used. Similarly, since a satellite can pass through the celestial pole, an equatorial mount is not best.

The Cloudcroft Observatory was designed to track space satellites although it has performed more classical astronomical tasks such as stellar photometry from time to time. It was developed and operated by the Air Force Avionics Laboratory. The mount has three axes, as shown in Fig. 2.11. The first, or azimuth axis, can be seen as the circle at the base of the telescope. Typically, the azimuth was positioned before a satellite pass and remained at one setting during the entire observation of that satellite. Usually, this axis was set to an azimuth which was midway between the rising azimuth of the satellite and the setting azimuth of the satellite. The second, or horizontal axis, can be seen in Fig. 2.11 as the round structure pointing out of the picture at the viewer. This axis was adjusted before a satellite pass so that the third, or tracking axis, was perpendicular to the orbital plane of the satellite. The third axis would then track the satellite along its orbit. Since the satellite's orbital plane is centered on the center of the Earth and Cloudcroft is located some distance from the center in New Mexico, minor adjustments were made in the second axis to precisely follow the object.[118]

This telescope was one of the earliest computer-controlled telescopes although control was never completely automatic. While the computer was used to set the axes prior to a satellite pass, human operators worked the controls manually to maintain telescope pointing and track the satellite. The computer was used to record telescope position in order to refine the satellite's orbital elements and to record photometric data. The telescope has been moved to Table Mountain Observatory, California, for more classical astronomical tasks.[119]

A different class of three-axis telescopes is the modified Dobsonian. Many observers have built inexpensive alt-az mounted, manually driven telescopes for reasons of cost. At a later time they want to take up astrophotography or perhaps they tire of constantly adjusting the telescope position while visually observing. This is particularly bothersome if the observer is trying to sketch. One obvious solution is to mount the whole telescope on a Poncet platform, as discussed in the section on Poncet mounts and other equatorial platforms.

The Optical Research Corporation of Appleton, Wisconsin, has taken a slightly different approach. They have modified the classical German-mounted portable

---

[118] A smaller version of this mount was used in the Moonwatch telescope described in *Sky & Telescope*, September, 1957, p. 535.

[119] The telescope is described in *Sky & Telescope*, February, 1980, p. 109. The Cloudcroft Observatory has a unique dome in that the slit opening continues through the top and down the backside. This allows the telescope to track satellites which pass through the zenith.

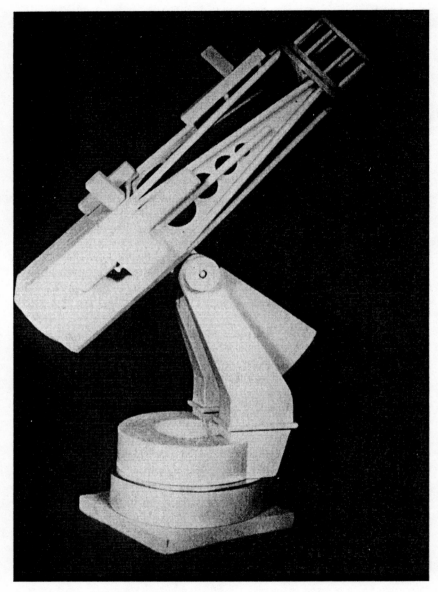

Figure 2.11. Model of Cloudcroft Observatory 1.22 m (48 in) telescope, photo courtesy of Ed Tyson.

Newtonian telescope and designed out many of the inconveniences. The mount features an alt-az assembly sitting on an equatorial table which can drive the telescope for about one hour. One obvious advantage of this system over a German mount is that the telescope tube cannot strike the pedestal. The designers borrowed some concepts from the Springfield mount and placed the altitude axis almost at eye level, as shown in Fig. 2.12. This means that as the telescope is moved from object to object, the eyepiece remains nearly stationary in height while the observer walks around the pedestal in azimuth. In this respect, operation of the instrument is similar to the Nasmyth telescope. The design also eliminated the need for rotating tube rings as seen on most German-mounted Newtonians. This results in a comfortable viewing position for any telescope orientation. The only mechanical disadvantage is that, like the Springfield mount, there is a large counterweight swinging around at just about eye level. A cautious observer might put a well-padded sock over the counterweights.

The designers of this telescope also modified the classical Newtonian concept by bringing the image to a focus inside the tube and then adding an achromatic relay lens to extract the rays. This has two effects; first the secondary mirror can be made much smaller, thus increasing the effective area of the primary mirror and decreasing diffraction effects. Second, the relay lens, which the designers have dubbed a Positive Transfer Element, can be varied in position between the secondary mirror and the eyepiece. This has the effect of changing the magnification of the overall system. Thus, with only two eyepieces, continuously variable zoom magnifications of 50 power to 675 power are possible.

Finally, the designers have added small features such as a curved secondary mirror support which minimizes the apparent diffraction effects and a finder scope which can be attached to the pier for easy polar alignment. Wooden inserts have been placed inside the top and the bottom of the pier tube to absorb vibrations. The telescope is also available in 20.32 cm (8 in) and 15.24 cm (6 in) versions.[120]

Four-axis telescopes

The commonest reason for four separate axes on a telescope is to allow the telescope to be moved, as in trailer-mounted telescopes discussed elsewhere. A second application is in Earth satellite tracking. As mentioned earlier, the path of a satellite across the sky can be approximated as a great circle and thus three-axis mounts are used. For precision tracking, however, a fourth axis is added to compensate for the fact that the satellite's motion is really a small circle on the sky. Precision tracking is required when small fields of view are used for photometry or imaging of the satellite. Typically, the two lower axes (azimuth and elevation) are manually operated and set before the pass of a satellite. The other two axes, typically

---

[120] The telescope is described in *Astronomy*, March, 1990, p. 60.

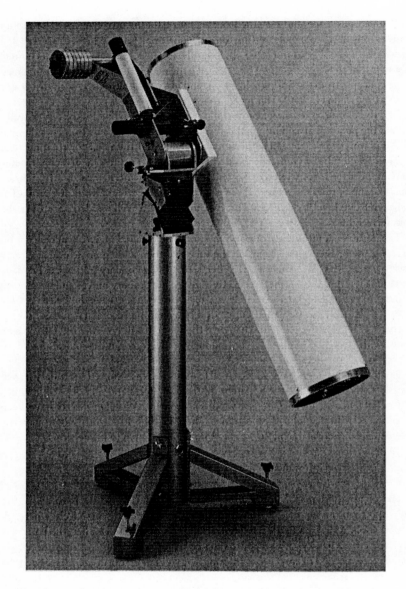

Figure 2.12. Three-axis 25.4 cm (10 in) f/5 telescope. Photo courtesy of the Optical Research Corporation.

Figure 2.13. Four-axis satellite-tracking telescope. The azimuth and elevation axes of this mount were originally used with a Baker–Nunn camera.

called the tracking axis and the cross-track axis, are then driven by motors controlled by either a computer or a human operator. The trailer-mounted telescope shown in Fig. 5.2 also uses this system.

The telescope shown in Fig. 2.13 was operated by the Air Force Avionics Laboratory in the 1970s. The main 61 cm (24 in) telescope fed a photometer used to measure the brightness of Earth satellites. The lower two axes were taken from a Baker–Nunn camera mount and the upper two axes were added later. It was co-located with the satellite search telescope shown in Fig. 1.15 near Wright-Patterson Air Force Base, Ohio. The two telescopes operated as a pair with the wide field search telescope locating satellites and passing on their coordinates to the four-axis tracking and photometry telescope.

Infinite-axis telescopes

Infinite-axis mounts are those which may move freely in any direction. Typically they are hand guided and the mount serves only to carry the weight of the telescope so that the observer doesn't have to hold it, although small hand-held rich-field telescopes might qualify as infinite-axis mounts.

During the age when refractors had only a single-element objective and focal lengths tended toward tens of meters, several infinite-axis mounts were designed.

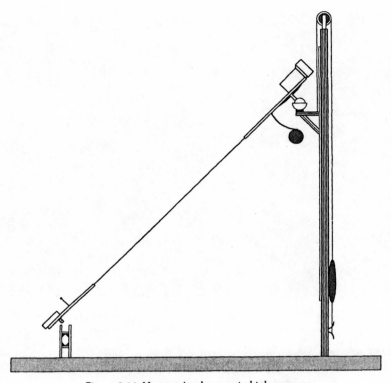

Figure 2.14. Huygens' pole-mounted telescope.

Typical among them was the pole-and-rope device built by Hevelius in the late 1600s, as shown in Fig. 8.3. At about the same time an instrument was constructed by Christiaan Huygens which alleviated some of the cumbersome properties of aerial telescopes, as they were then called.

Huygens' telescope was supported by a tall pole with an elevator platform controlled by a rope. The platform carried a spherical mount for the objective lens assembly which was mounted on a stiff rod, as shown in Fig. 2.14. The rod was attached to the eyepiece by a rope. By keeping the rope taut, the eyepiece and objective, separated by 37.5 m (123 ft), were automatically aligned with each other. The observer held the eyepiece assembly and aimed the telescope manually. A movable wooden support provided a resting place for the observer's arms, which might grow weary from holding the eyepiece assembly and rod.

While this arrangement eliminated many of the beams and spars of Hevelius' design, it became difficult to point the telescope on very dark nights when the objective assembly could not be seen. At times a lantern was needed to determine where the other end of the telescope pointed. Huygens' telescope was reputed to have been less susceptible to wind gusts and vibrations than Hevelius'.[121] When observing stars on the celestial equator, Huygens had to move the eyepiece about 16.6 cm (6.4 in) every minute. At high magnifications, this requires a steady hand and the innate ability to walk around a pole at sidereal rate.[122]

While Huygens' telescope may be classed with pole-and-rope mounts, the basic mechanical pivot was the ball in a cup on the elevated platform. Spherical mounts have an admirable history. The first Newtonian telescope sat on a ball held in place by wrought iron springs. Good high quality spheres capable of supporting the weight of a telescope can be expensive if manufactured expressly for a mount. However, telescope designers are experts at making do with available materials and this is the origin of the popular bowling ball mount.[123] A typical application is shown in Fig. 4.8.

One advantage of spherical mounts for hand-guided telescopes is that there is no dead zone where one axis must move infinitely fast in order to track the stars, as is the case with alt-az mounts. Spherical-bearing mounts have been used on a variety of telescopes with apertures of up to 1.5 m (60 in).

Floating telescopes

Often telescopes are said to float upon thin oil-film bearings. These are hydrostatic oil films used to separate the rotating components from the static ones. At least one telescope, the Kuiper Airborne Observatory 0.9 m (36 in) instrument, uses a high pressure air bearing to isolate the optical system from the vibrations of the aircraft which carries it.[124]

Telescopes have, however, been built which float by displacement of fluid in much the same way that a boat floats on the water. The Common Telescope at Harvard Observatory in Cambridge, Massachusetts, had a right ascension axle which floated ⸍ in a tank of water, as shown in Fig. 2.15.[125] Since the main weight of the telescope was not on the two bearings, smaller bearings could be used and much smaller forces were required to move the telescope. The telescope is also noteworthy in that it is a

---

[121] Huygens' telescope is described in *The History of the Telescope*, Henry C. King, Dover Publications, 1955, p. 55.

[122] A modern version of this telescope with a focal length of 13.7 m (45 ft) was built by Chuck Thompson and shown at the Illinois Astrofest. It is described in *Telescope Making*, No. 25, Winter, 1984, p. 18.

[123] Mountings using bowling balls can be found in *Telescope Making*, No. 8, Summer, 1980, p. 12 and No. 34, Fall, 1988, p. 35. See also *Sky & Telescope*, April, 1957, p. 293, June, 1972, p. 368, October, 1972, p. 237, March, 1974, p. 172, August, 1980, p. 160 and August, 1985, p. 473.

[124] An eight meter aperture telescope has been proposed in which the telescope is surrounded by a twenty meter diameter steel sphere. The sphere is supported by air pressure much like an egg in an egg cup. The sphere also forms the dome of the telescope structure, thus minimizing the cost of the structure. See the paper by Horace W. Babcock, *Proceedings of the European Southern Observatory Conference on Very Large Telescopes* (1,99), Garching bei München, 21 March 1988.

[125] *Sky & Telescope*, January, 1978, p. 20. See also *The History of the Telescope*, Henry C. King, Dover Publications, 1955, p. 276.

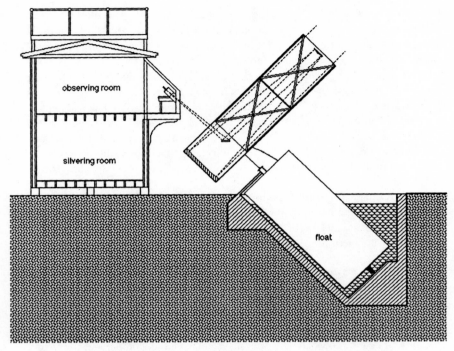

Figure 2.15. Common Telescope at Harvard Observatory. Drawing is copied from an illustration courtesy of Harvard Observatory.

coudé with the observer sitting in a warm building adjacent to the telescope itself. One wonders how the pool of water was kept from freezing in the harsh New England Winters. A new mounting was made for the mirror in 1933 and the mirror was later refigured.[126]

Other floating telescopes are the Mount Wilson Observatory 1.52 m (60 in) and 2.54 m (100 in) which have moving parts floating in collars of mercury.[127] Since the mercury is totally encased in the mount, both of these telescopes look fairly normal, as if they had conventional roller bearings. Similarly, the siderostat base of the Great Paris Refractor of 1900, shown in Fig. 8.1, floated in a pan of mercury.

[126] *The History of the Telescope*, Henry C. King, Dover Publications, 1955, p. 395.
[127] *Sky & Telescope*, January, 1976, p. 13. See also *Sky & Telescope*, February, 1976, p. 95.

3

Telescope and mount materials

Ostensibly, the purpose of the telescope tube is to hold the optics together. Occasionally, it will be used to block stray light. At times, however, the shape of the instrument is designed more to please the eye than to be part of an optical system. While this often results in severe compromises, a notable exception is the Porter Garden Telescope.[128] Russell W. Porter was always looking for ways to get people interested in astronomy and turning the telescope into an attractive garden ornament seemed ideal. The 15.24 cm (6 in) aperture Newtonian telescope, shown in Fig. 3.1, was cast in bronze. Eyepieces of 25 power, 50 power and 100 power were supplied.

The instrument was fitted with a mirror cover and was designed to be left in the garden in all weather. It doubles as a Sundial. The castings depicted a lotus bowl holding the primary mirror with one long leaf holding the eyepiece. Control knobs are shaped like small flowers. The mount is actually a three-axis type with azimuth, right ascension and declination axles. During the day time, the local hour angle is set to the meridian and the polar axis is clamped with the declination axis horizontal, acting as an elevation axis. The azimuth axis is then unclamped and the telescope can be used to observe terrestrial objects as an alt-az telescope. At night, the azimuth axis is rotated until the polar axis points at the pole and then the azimuth axis is clamped. The polar axis is unclamped and the telescope is used as a classical equatorial mounted instrument. At least fifty of the telescopes were made from the castings and several survive today.

Several years ago I made the statement that if a material exists, it'll be used in a telescope. I was challenged almost immediately by an obstreperous astronomer to find a telescope structure made of half-cooked spaghetti. A few weeks later I visited Mike Reed who was working on the preliminary mechanical design of the Fred Lawrence Whipple Multiple Mirror Telescope in Tucson, Arizona. The entire structure of the telescope had been modeled in a computer in order to determine the minute flexures expected in the telescope. As the telescope points to different areas of the sky, the entire structure bends and sags enough to throw the six main mirrors out of alignment with each other. It was reasoned that you could never prevent materials from bending and sagging but you could at least compensate for it, as in a

---

[128] A history of Porter's 1923 design can be found in his biography, *Russell W. Porter* by Berton C. Willard, Bond Wheelright Publishers, 1976, distributed by Sky Publishing, Inc.

Figure 3.1. Porter's Garden Telescope. Photo courtesy of John F. Martin, V.

Serrurier truss. A great deal of engineering went into the design of the optical supports in order to compensate exactly for the bend and sag. Sophisticated computers actively control the mirror pointing. The structure does flex but the computers compensate for the sag. "It was," Mike said, "like trying to build a telescope out of half-cooked spaghetti."

The usual method of telescope design is to try to make the tube as rigid as possible within the limits of mass and cost. The commonest tube is a cylinder but there is a stronger structure and that is the truncated cone. Olaf Römer in 1684 constructed one of the first telescopes with a truncated-cone tube, also called a double-cone tube.[129] The telescope made for the Reverend John Craig and shown in Fig. 2.5 is another example. In more modern times the truncated cone has been used to make

[129] *The History of the Telescope*, Henry C. King, Dover Publications, 1955, p. 104.

lightweight telescopes with low moments of inertia for fast slewing. The satellite search telescope shown in Fig. 1.15 is a typical example. With its riveted aluminum stressed-skin monocoque construction, the telescope is built more like an aircraft fuselage than a telescope tube.

Tube materials

Wood is the commonest telescope material but it is seldom used for the tube itself. This is because tubes are traditionally round and commercially prepared wood comes in flat pieces. Many people have made square or multi-sided wooden telescope tubes but in the modern era few have made round tubes.[130] Other materials are much more versatile, although they lack the esthetic qualities of wood, which is discussed in detail later in this section.

Ceramics have been used as tube materials in several telescopes. While this makes for a heavy tube, it is of little consequence if the tube does not have to move. The Tokyo Observatory operates a zenith telescope which requires high thermal stability in the tube. The telescope makes timings of star passages over the observatory each night and the readings from night to night are compared. It is thus important that the telescope remain constant in focal length and pointing position as the temperature varies. Porcelain, with its low coefficient of thermal expansion was selected for the telescope structure, as shown in Fig. 3.2.[131]

Instead of trying to make a telescope which doesn't change with temperature, the designers of the McMath solar telescope at Kitt Peak made a telescope tube which controls the temperature. Since the telescope tube is 152 m (500 ft) long, the engineers were worried that thermal air currents could form inside the tube and degrade the seeing. The telescope, shown in Fig. 7.22, is encased in a building whose walls are copper, a metal which conducts heat well. The copper walls have pipes embedded in them through which heated or chilled water is pumped in order to maintain a constant temperature throughout the structure. In order to further control daily temperature swings, about half of the telescope tube is buried below ground level. Thus, there is a constant temperature maintained throughout the long optical path, resulting in high resolution images of the Sun.

Concrete has been used as a telescope tube material. While weight is usually a design consideration and concrete is not known for its lightness, there are applications where it works well. The CERGA interferometer telescopes, located in the French Alps, required spherical shells about 3.5 m in diameter for the lower tube assembly. Each telescope base, shown in Fig. 4.13, mates with a unique support ring which drives the telescope sidereally. When faced with a choice of materials to produce the shells the designer, Antoine Labeyrie chose concrete. Metal curved surfaces would probably have required expensive machining. The concrete was finished with a smooth surface which conforms to the base-ring support.

---

[130] A notable exception is shown in Don Raether and Carl Blue's telescope, described in *Telescope Making*, No. 21, Winter, 1983, p. 20.   [131] *Sky & Telescope*, November, 1975, p. 276.

Figure 3.2. Ceramic tube zenith telescope. Photograph courtesy of Yoshihide Kozai, National Astronomical Observatory, Mitaka, Tokyo.

Wood

Wood has an honorable tradition in telescope design. America's largest telescope from 1847 to 1863 was the 38.1 cm (15 in) refractor at Harvard Observatory. This wooden-tube instrument was used for years and made such discoveries as the crepe ring of Saturn.[132] Although wood tubes change length with temperature and they swell with moisture, proper finishing of the surfaces minimizes these problems. With the advent of more modern materials such as aluminum, plastics and composites, the use of wood has been relegated to the amateur telescope designer.

The highly polished wood surfaces of Weyman Reams' telescope, seen at the Riverside Telescope Makers Conference in 1981, make the telescope look more like

[132] The telescope is described in *Sky & Telescope*, January, 1977, p. 14.

Figure 3.3. Bent wood telescope made by Weyman Reams.

furniture than the scientific instrument it really is. The hand-crafted bent wood mount, shown in Fig. 3.3, took many years to make. The wood was soaked in a special brine until it became soft and was then bent into the desired shape. Layers of curved boards were then laminated to make the final structural members. The tube has 12 sides and is made of mahogany. In addition to looking good, the 20.32 cm (8 in) f/8.3 telescope performs well optically and is easy to use.[133]

One of the longer lived unusual telescopes has a wooden lattice tube, as shown in Fig. 3.4. Originally known as the Grubb Reflector, it was built in 1862 and mounted in Melbourne, Australia in about 1868. The 1.21 m (48 in) aperture mirror was the last large reflector to use speculum metal for the mirror.[134] It is interesting to note that prior to this time, telescopes were identified by their focal length instead of their aperture. This telescope became known as the Four Foot Melbourne Reflector or The Great Melbourne Reflector and the practice of listing apertures has continued right up to the 6 m (236 in) telescope. A lattice tube is stronger than the traditional eight-posted Serrurier truss and yet it does not have the wind loading of a solid tube. The telescope was later modernized, a glass mirror replaced the original and it has survived into the modern era at Mount Stromlo Observatory.[135]

[133] A description of the telescope can be found in *Telescope Making*, No. 12, Summer, 1981, p. 26.
[134] *The History of the Telescope*, Henry C. King, Dover Publications, 1955, pp. 261–7 & 392. See also *Astronomy*, Fred Hoyle, Crescent Books, 1962, p. 55.
[135] *Sky & Telescope*, June, 1956, p. 342.

Figure 3.4. The Great Melbourne Reflector. Notice the lattice tube construction.

Mount materials

To the telescope maker, any spare material not nailed down or on fire is fair game for conversion into telescope parts.[136] Items that are nailed down may require the application of a hammer or crowbar before being converted. When chasing eclipses and occultations overseas, I often build a telescope mount from lightweight materials simply to save the cost of shipping. This concept can also be seen in Ernie Piini's telescope, as shown in Fig. 1.42 and in Fig. 3.8. I design my mount to use some local heavy materials such as rocks or sand for counterweights and base supports. On one

[136] In *Telescope Making*, No. 6, Winter 1979/1980, p. 48, Reg Spry of Sussex, England describes a telescope built of "two shelves out of the larder cupboard, half-shaft of a car, part of a car jack, bits of Meccano™, plastic tops of coffee jars and any other old scrap lying around". And this is for a telescope that's not particularly unusual in outward appearance. A complete description of his observatory is in *Telescope Making*, No. 8, Summer, 1980, p. 30.

occultation trip to Ecuador, I cast about for a likely counterweight rock and found the perfect specimen. My host became quite agitated, thinking that I was going to keep the rock. Only later did I discover that he had been saving the rock, a large semi-precious stone, to cut and polish. By now my counterweight is a set of necklaces, earrings and assorted jewelry.

One of the more unusual examples of unique mount material was constructed by C. D. Rorke. It is for a 15.24 cm (6 in) mount using 5,812 Meccano™ (Erector Set™) parts.[137] The yoke-mounted telescope includes a clock drive.

Synthetic materials have been applied to telescopes successfully. While in the past, telescope designers were limited to wood, brass, aluminum and steel, today's telescope creator has a wide variety of materials and methods of fabrication such as the composite material telescope shown in Fig. 1.3. One example is "Gatorfoam™". The material is made in slabs from 3/16 inch to 2 inches thick. It is composed of a core of Styrofoam® faced with plasticized long-fiber birch paper which resists dew. This makes for a strong, lightweight structural material. One of the first telescopes built from this material was made for a child, so carrying weight became important. The instrument, built by Bob Kelly of Columbus, Indiana, has withstood years of observations.[138]

Poncet mounts and other equatorial platforms

Most equatorial mounts are designed so that they could track a star from horizon to horizon over a twelve-hour exposure. Indeed, some telescopes beyond the Arctic or Antarctic circles are capable of continuous operation during the long dark winter. Most observations do not last that long, however. The majority of visual observations last only a few minutes before moving on to the next object. Why, then, do many large telescopes have a complete right ascension gear when only a small portion of it will be used on any given night? Large worm-wheels are expensive. Why not build the telescope to move equatorially for, at most, an hour or two?

This is the basis of Adrien Poncet's wonderful invention. While equatorial platforms had been built before, Adrien is responsible for reducing the concept to practice and showing how many amateurs could, with simple tools, make their own. Basically, a Poncet mount is a small section of a horseshoe mount.[139] It is

---

[137] *Sky & Telescope*, June, 1971, p. 384.

[138] The telescope is described in *Telescope Making*, No. 13, Fall, 1981, p. 5. Gatorfoam™ is produced by the International Paper Company of Statesville, North Carolina.

[139] Poncet mounts are discussed in a special issue of *Telescope Making* devoted to Adrien Poncet, issue No. 14, Winter, 1981/1982. See also issues No. 8, Summer, 1980, p. 22, No. 10, Winter, 1980/1981, p. 12, No. 25, Winter, 1984, p. 9 and No. 37, Summer, 1989, p. 41. The original Poncet mount is shown in *Sky & Telescope*, January, 1977, p. 64. See also the issues of November, 1985, p. 489, February, 1980, p. 163 and March, 1980, p. 251.

Figure 3.5. Detail of a Poncet mount mechanism beneath an observatory. Photograph courtesy of Jack Ells.

characterized by having a low height with the telescope, usually an alt-az instrument, placed on top of the platform.

An entire observatory has been constructed on a Poncet mount, as shown in Fig. 3.5 and Fig. 3.6. J.W. Ells of Bexley, England, built a small hut on a platform designed by E.G. Hill and mounted his 32 cm aperture Newtonian in an alt-az configuration within. In the photograph above, the stub polar axle can be seen at the center of the framework. The small plates around the periphery support the weight of the observatory building and are all perpendicular to the polar axle. Small wheels ride on the plates as the structure rotates about the axle.

The hut keeps members of the Crayford Manor House Astronomical Society warm while observing.[140] The hut is heated but this doesn't affect the seeing since the observer's chamber is insulated from the telescope area. Only the eyepiece assembly penetrates the wall separating the observer from the telescope. The telescope itself moves only in elevation with respect to the observing hut while the entire hut can move in azimuth. The observatory is used for planetary photography and photoelectric photometry.

In a related development, a box-and-hinge mount was constructed by Tom

---

[140] *Journal of the British Astronomical Association*, Vol. 89, No. 1, December, 1978 p. 66. See also *Sky & Telescope*, March, 1980, p. 251.

Figure 3.6. Observing hut mounted on a Poncet platform. Photograph courtesy of Jack Ells.

Fangrow. This design, shown in Fig. 4.12, consists of a cubical box sliced diagonally. A hinge connects the two sections together. When setting up the telescope, the hinge pin is pointed at the pole star and the platform then tracks the stars. While taller than a classical Poncet mount, the design is a bit easier to build.

Axle materials and bearings

Axles are one of the more difficult parts of a telescope to manufacture by hand (once you've ground the mirror). Telescope tubes, mounts and even eyepiece holders can be made using rough hand tools and a little effort. Axles, on the other hand, must rotate smoothly and that implies precision machining. They must also support the telescope and this implies strength. One exception to this is the Dobsonian mount azimuth axle which depends on a bolt for centering and the particular friction of Teflon® on Formica® for a bearing surface. In general, however, most telescope builders must either buy a manufactured set of bearings or borrow parts from some other mechanism.

A farm lad in the 1920s near Burdett, Kansas, built a 22.86 cm (9 in) f/8.7 telescope by hand and made the mount out of parts from a 1910 Buick, a cream-separator base and mechanical components from a straw spreader. Economics forced him to tailor

Figure 3.7. Newtonian telescope made from farm machinery. Photo courtesy of Dr Clyde Tombaugh.

his telescope design to fit the available scrap parts. While discarded farm machinery has been used for many an amateur telescope, this one was special. More than half a century later it was still working in his back yard, providing excellent views. In the meanwhile, he used the telescope to make some planetary sketches, sent them off to Lowell Observatory and so impressed the professional astronomers that he landed a job there. Ultimately, that hand-built telescope allowed him to have access to more professional instruments, resulting in the discovery of Pluto. Clyde Tombaugh's telescope design was born out of the necessity of using inexpensive available materials but he did not scrimp on the quality of the instrument.

Its mirror was hand-ground and tested in a storm cellar. It's not the most elegant looking optical instrument I've ever used but it is one of the better planetary telescopes around. This telescope, shown in Fig. 3.7, and several other hand-built

Figure 3.8. Using a telescope packing case as a mount during a solar eclipse expedition in Kenya.

instruments adorn the yard at Dr Tombaugh's retirement residence. Because of its role in the the history of astronomy, I would classify this as one of the more important telescopes in the world.[141]

Mount vibrations

In addition to smoothly pointing to stars, the telescope mount must be immune to vibrations and motions of the ground and structures surrounding the telescope. While it is a common practice to isolate the telescope support from the observatory building structure and dome, this is not always feasible. In addition, there are some soils such as caliche clay which transmit vibrations from nearby sources such as trucks on an adjacent roadway. My own telescope sits on caliche and a heavy vehicle passing my home can ruin a good astrophoto. Maybe that's why I gave up astrophotography and concentrated on video astronomy. Experiments with sand-filled steel pipe piers helped but the problem persists.

In portable astronomy I have used the technique of making the telescope case into

[141] A description of the telescope is given in *Telescope Making*, No. 23, Summer, 1984, p. 4. The history of this instrument is detailed in Tombaugh's book *Out of the Darkness, the Planet Pluto*, written with Patrick Moore, Stackpole Books, 1980.

a pier. For overseas eclipse observations I didn't want to pay the freight costs on the telescope pier or tripod so I made an adapter plate for the equatorial head which bolts on the telescope's packing case. The result, shown in Fig. 3.8, is a delicate but serviceable mount. The mount becomes much more steady when plastic garbage bags filled with locally available sand or soil are placed in the case before attaching the telescope. It's a massive mount without the expense of shipping a hundred kilograms or so of telescope mount materials by air. A similar technique was used by Jan Hers in South Africa to stabilize a telescope used in seeing tests. He made the mount from an old water tank which could be emptied when the telescope had to be moved.[142]

In some cases local vibrations cannot be avoided. The solution may lie in complete mechanical decoupling of the telescope from the ground or building. Although this is contrary to most generally accepted practices of mounting the telescope sturdily to the Earth, it has been shown to work in some cases. For instance, at Arizona State University the student telescopes are situated on the top of a classroom building several stories tall. Adjacent to the rooftop observing area is the building's air conditioning plant with heavy motors and fans. When the fans and compressors run, you can feel the vibrations through the soles of your shoes while observing. The solution is not to turn off the air conditioner during observations since it is often hotter than 38 °C (100 °F) at midnight in Tempe, Arizona. An ingenious solution was designed after considerable experimentation by students. Each leg of the school's Celestron C-14 telescope tripod is placed in a coffee can which has been cut down to form a holder for three tennis balls. The tripod leg tip nestles in between the tennis balls. The telescope wobbles a bit when bumped or slewed but the vibrations dampen out after a couple of seconds. The vibrations from the air conditioner are not visible even when high magnification planetary viewing is performed. Recently commercial manufacturers have started selling vibration-damping pads.[143] These are composed of a shallow metal cup filled with a rubbery compound such as silicone adhesive. In the center of the mass of adhesive, isolated from the rim of the cup, is a small metal pad which accepts the telescope tripod leg.

A second and more elaborate approach was developed by Andrew Wagner and Phillip J. Stiles at Brown University in Providence, Rhode Island, for a similar application of a student telescope on top of a classroom building. As at Arizona State University, the telescope must be situated on top of a tall building to elevate it above the glare of nearby streetlights. The solution at Brown was to build a heavy cubical frame anchored to the floor, as shown in Fig. 3.9. Within the frame is a massive box hung by springs from the outer frame. The telescope is attached to the weighted box and it projects out over the side of the frame. Counterweights in the box balance the floating mass of the telescope. Care was taken in the design to support the combined mass of the box and the telescope below its center of mass, making it slightly top

---

[142] *Sky & Telescope*, September, 1972, p. 194.
[143] *Proceedings of the Riverside Telescope Makers Conference*, May 26–29, 1989, p. 31.

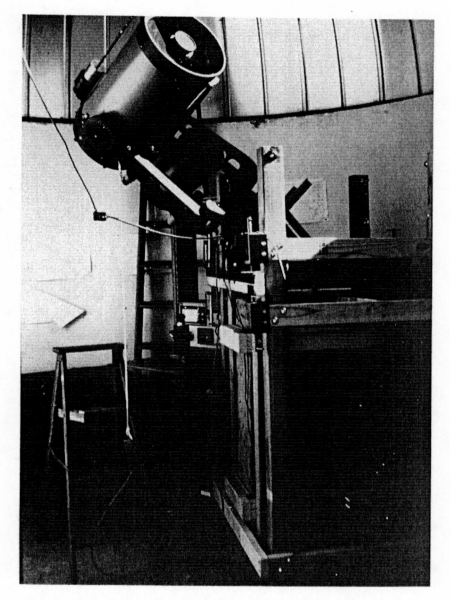

Figure 3.9. Mount designed to dampen building vibrations. Photo courtesy of Brown University.

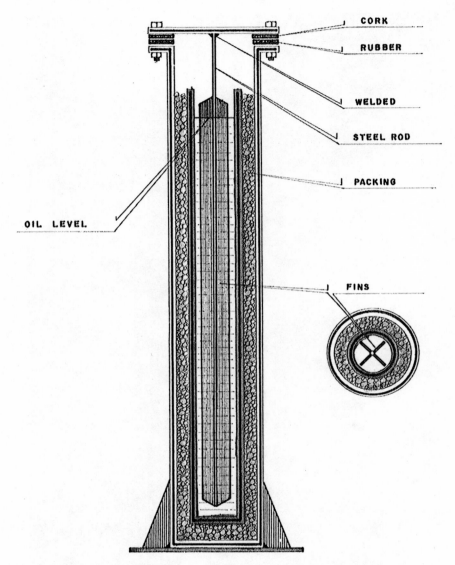

Figure 3.10. Springless-damper oil-filled mount. Drawing by Rafael López Vélez.

heavy. This resulted in telescope vibrations that are translational but not rotational. Rotational vibrations will show up as pointing errors but translational vibrations are permissible since they are not visible as an image shift in the eyepiece.[144]

Rafael López Vélez of Mexico solved the rooftop vibration problem by filling a tall vertical pipe pier with heavy oil. The plate at the top of the pier was then decoupled from the pier by insulating it from the pipe with rubber and cork gaskets. A steel rod with fins was then attached to the bottom of the plane. The rod extended down into the oil and the fins coupled the plate to the viscous oil, thus dampening vibrations, as shown in Fig. 3.10. This technique is especially effective in minimizing rotational vibrations of the telescope while ignoring translational vibrations which do not interfere with observations.[145]

The trick in ridding a telescope of vibrations is to determine first whether the vibrations are caused by high frequency sources such as motors and fans operating at 30 to 100 Hz or low frequency sources such as people walking in the building. Higher frequency vibrations can be nullified with dampening mechanisms such as the silicone-filled pads which fit under the tripod's legs. Lower frequency vibrations require more elaborate devices such as the one developed by Brown University. If, however, you live in California and the very ground you stand on shakes, then the solution is to move to Arizona.

[144] *Sky & Telescope*, September, 1987, p. 314.   [145] *Telescope Making*, No. 34, Fall, 1988, p. 20.

4

Strange drivers

This chapter is devoted to all of those mechanical and electrical engineers who have worked long and hard simply to counter the Earth's rotation. The obvious reason for driving the telescope is to track the stars. At first glance, it would seem an easy task to turn a telescope axle at one revolution per day. Two major problems occur; the first major problem is that the *solar* rate is one revolution per day. The stars, at *sidereal* rate, are about four minutes per day different. Thus, a common 60 Hz (50 Hz in Europe) synchronous electric motor and a simple gear train will not exactly track the stars. Two solutions have been tried; the first is to make the motor run a little slower, which requires an electronic drive corrector. This is a source of AC power that runs at slightly less than the 60 Hz of the power company. The second is to develop a gear train that deviates just a little from the solar rate. The first solution requires a bit of electronics expertise (or money) and the second requires at least 11 gears. Drive correctors are the commonest solution, although spring-driven and weight-driven clocks, computer-controlled stepping motors and a variety of other methods have been tried.

J. Van Raalten of de Bilt, Holland constructed a drive with six special gears from a Meccano™ set (the European version of the Erector Set™) plus a motor geared down to 1 rpm.[146] The result, shown in Fig. 4.1, worked well but his problem was that the power company frequency wasn't stable and the tracking drifted. He added a second motor whose output was summed with the main driving motor and which can guide the telescope manually, compensating for the power company.

In most developed countries, the power company is required to assure that there are exactly 5,184,000 cycles of power per day (4,320,000 in Europe and other 50 Hz countries). This is so that all the electric clocks in homes and offices remain accurate. Power companies are allowed to deviate a little but they must make up for a slow or fast generator by compensating later. It is easiest to tweak the rate on a generator when there isn't much electrical load on it and thus most frequency variations will occur at night. Unfortunately, this is the very time when astronomers need stable power.

The second major problem in driving a telescope smoothly is that even if you can

---

[146] The drive is described in *Sky & Telescope*, September, 1973, p. 185.

Figure 4.1. Drive corrector built from a Meccano™ set. Photo courtesy of J. van Raalten.

get a motor shaft to turn at exactly the correct rate, the reduction gears will distort that rate. For centuries man has tried to make a perfect gear in which the rotation rate of the driven gear is exactly proportional to the driving gear. The problem is that the gear teeth engage first at the top of the tooth and then farther down the tooth and finally at the root of the tooth. Consequently, the radius of the gear changes as a function of where the gear teeth are engaging. The speed ratio of two gears is only approximately proportional to the ratio of the number of teeth in the gears. It is exactly proportional to the ratio of the radii, from the center of the gears to the point of contact on the teeth. This can be minimized by using worm gears and worm wheels which have a lesser gear error.

Even the best of telescopes will track stars first a little fast, then a little slow, then a little fast. This is called the periodic gear error and its greatest component will be determined by the last driven gear, usually the right ascension worm gear. The total accuracy of the drive is thus a function of both the drive motor accuracy and the gear accuracy. It is pointless to use a quartz crystal controlled drive corrector which is accurate to one part in ten million in order to drive a spur gear which has errors of one part in a hundred. For most visual observers, having the stars wobble a few arc seconds every three or four minutes of time will not be a problem. For astrophotographers, however, the situation is intolerable. Just about the only cure is to vary the

Figure 4.2. Periodic gear error corrector. Photo courtesy of Louis J. Faix.

speed of the driving motor to correct for gear problems. This usually requires that the observer monitor the gear error visually with a guiding eyepiece and vary the speed of the drive motor manually via the drive-corrector controls.

Some rather unusual control systems have been invented to eliminate the periodic gear error. Louis J. Faix of Washington, Michigan corrects for the error in a worm gear driving the final polar gear by linking the worm gear to an electrical potentiometer which controls the rate of his drive corrector. Since most of the periodic gear error is due to the mesh of the worm gear and the final gear, most of the tracking error will show up at the same period of revolution as the worm gear driving the final gear. The worm gear has an eccentric attached to it as shown in Fig. 4.2. The eccentric rod pushes a lever which rotates the shaft on the potentiometer. Considerable adjustment is required to match the amplitude and phase of the correction to the amplitude and phase of the gear error. The method requires precise modelling of the errors of both the worm gear and the worm wheel (polar axis gear) but the results are very good.[147] The mechanical cams and linkages required, however, are probably not for the beginner. After correcting the periodic gear error, Faix discovered that lesser errors in the gears were also present. The worm wheel itself had tooth to tooth variations. Lapping the worm wheel and worm gear with

[147] A description of the drive corrector is given in *Sky & Telescope*, May, 1978, p. 439.

light oil and fine abrasive decreased that problem. Precise alignment (perpendicularity) of all gears is important, as this affects the pressure between gears. Finally, after achieving stable long-term tracking, minute dust particles crept into the gearbox and threw the calibration off.

Faix's electro-mechanical solutions to the problems of drive errors is probably near the limit of what can be done without actively sensing any image position drift and correcting it in near real time. Such a practice, commonly called guiding an astrophoto, is tedious and boring work. Considerable development has gone into automating this task. Engineers are lazy, if not ingenious. While spending hours at an eyepiece guiding astrophotos, many have dreamt of building an automatic star guider which will eliminate this tedium. It isn't very difficult to sense the star position and feed the signal back into the drive corrector. It has been done on professional telescopes for years. One of the earlier references to an automated telescope guider was on the Harvard Observatory coronagraph solar telescope drive at Climax, Colorado.[148]

Amateurs, however, have not produced many automatic-guider designs. Did you ever wonder why so few automatic star-guider designs ever appeared in print? The technology is available and the engineers are familiar with it. The required equipment is cheap enough and there aren't any insurmountable problems that haven't been dealt with by dozens of professionals. The problem lies in the way amateur astronomers pursue astronomy. As a warning to those inclined to make such a design effort, I must relate the story of at least two astronomy clubs that have undertaken such projects, the San Jose Astronomical Association and the Saguaro Astronomy Club of Metropolitan Phoenix, Arizona. Both of these organizations had members who were professional electrical and mechanical engineers, capable of successfully designing a guider. The problem was that the engineers became so interested in photodetector design, television systems, computer interfaces and software that they never took another astrophoto. The true photographers either bought a commercially available automated star guider or went back to hand-guiding long exposures while muttering about those crazy electron pushers who give astronomy a bad name.

One method of minimizing the periodic gear error is to make the final driven gear (the one attached to the right ascension axle) as large and as accurate as possible. This implies that a deviation in the spur or worm gear driving the final gear will have a lesser effect than if the final gear were smaller. Thus, we find large brass worm gears a meter or more in diameter on some professional telescopes. Alas, these are expensive pieces of machinery. Some clever engineering has been done to get around these limitations such as the technique of fitting a bent rack gear to a large diameter disk.[149]

Louis Boyd of the Phoenix Astronomical Society, in one of the first fully automated telescopes, used a large aluminum disk with a motorcycle chain wrapped

[148] *Sky & Telescope*, February, 1945, p. 4. See also *The History of the Telescope*, Henry C. King, Dover Publications, 1955, p. 380.   [149] *Telescope Making*, No. 33, Summer, 1988 p. 18.

Figure 4.3. Louis Boyd's fully automated telescope. Note the large disks with motorcycle chains wrapped around them which form the drive system.

around the periphery, as seen in Fig. 4.3. Since it's much cheaper to produce a smooth disk than a disk with gear teeth, this saves considerable expense. In Boyd's case, he found the disks already fabricated at a surplus-metal yard. The 25.4 cm (10 in) aperture telescope was designed from the start for computer-controlled unattended photometry of stars. No human observer operates the telescope. Thus, it does not have a regular eyepiece but instead it has a permanently mounted photometer. The telescope, originally built in Phoenix, Arizona, has been moved to Mount Hopkins, Arizona, in order to enjoy the darker skies. It has become part of the Automated Photoelectric Telescope Service.[150]

The motorcycle chain has been replaced by a steel band or cable in several telescopes.[151] The smooth action of a steel band eliminates any possibility of

[150] Boyd's telescope is described in *Telescope Making*, No. 22, Spring, 1984, p. 28. The telescope is fully automated and it observes on clear nights unattended. I have had the honor of watching the telescope go through its paces, making photometric measurements in several colors, finding the comparison star, making measurements, checking the sky background, moving on to the next variable star and repeating the whole process until dawn. Frankly, after about five minutes, the process is rather boring to watch so Lou and I usually drift off to chat about new computer systems.

[151] A steel band drive is described in an article in *Telescope Making*, No. 11, Spring, 1981, pp. 37 f. The drive is described on p. 42. A second steel band drive by Tom Pock using a tape measure is described in *Telescope Making*, No. 34, Fall, 1988, p. 18.

nonlinear motion caused by "cogging" of the motorcycle chain. Usually, the steel band or cable is not attached to the driven disk and relies on friction to keep the right ascension axle from slipping, as in Boyd's telescope. In some installations the band is clamped to the driven disk.

Dr Frank Melsheimer of DFM Engineering went one step farther by eliminating the chains and simply driving the disk with a small smooth roller. While others have done this before and after his work, he has optimized the engineering of the hardness of the roller, the ratio of the size of the disk to the roller and the the support of the entire assembly. He now produces the drives commercially.

Tangent-arm drives

Making the final driven gear large spawned the idea of the tangent-arm drive, where the final gear is an incomplete and flattened sector of the whole RA gear. Typically, it is driven with a threaded push-rod which engages a threaded collar on the arm, as seen in Fig. 1.14 and Fig. 1.28. The problem with a tangent-arm drive is that near the ends of its travel, its rate is not the same as at the center. André Hamon solved this problem by connecting the lead screw and the tangent arm through a special cam arrangement.[152] The collar which holds the lead screw is shaped to push the tangent arm at a constant angular rate even though the collar moves at a nonlinear angular rate as measured from the center of rotation of the arm. The Hamon cam drive has been used successfully on Steve Dodson's telescope, shown in Fig. 5.5.

If you don't like that solution then the cure is the old curved-rod trick which is similar to the curved-bolt drive shown in Fig. 4.7. In this variation of the tangent-arm drive, the threaded rod is not rotated but the threaded collar is. Then the rod is bent to a radius equal to the distance between the axle and the driven collar. The rod is then clamped at both ends.[153] The advantage here is that a threaded rod, carefully bent by hand, has about the same gear pitch accuracy as a conventional worm gear of comparable radius but at a fraction of the cost. In fairness to all gear cutters (a job possibly more daunting than grinding mirrors) the threaded rod doesn't have the strength of a good gear and it must be reset at the end of its travel.

The tangent-arm drive and the curved-rod idea have been combined to produce a sector of a worm wheel by casting threads into the curved end of a tangent arm. Typical materials used to make the cast threads are wood putty[154] and epoxy-filled resin.[155] An example of the type by A. L. Woods is shown in Fig. 1.29. A second

---

[152] The original article by André Hamon is in *Sky & Telescope*, June, 1978, p. 531. One application showing the engineering details is in *Telescope Making*, No. 21, Winter, 1983, p. 50.

[153] A good example of a tangent-arm curved-bolt drive by Jack Roach and Kevin Hotten is seen in *Telescope Making*, No. 17, Fall, 1982, p. 29. A similar drive is also used on the declination axle of the telescope. A complete description of the telescope is given in *Telescope Making*, No. 20, Summer/Fall, 1983, p. 22.   [154] *Sky & Telescope*, April, 1979, p. 392.

[155] A description of how to cast epoxy worm gear teeth is given by A. L. Woods in *Telescope Making*, No. 30, Summer, 1987, p. 14. See also *Sky & Telescope*, October, 1979, p. 306.

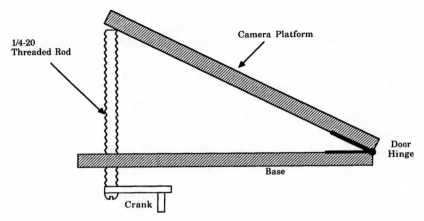

Figure 4.4. Scottish mount diagram. A wide field camera may be mounted on the upper board. If the crank is turned at one revolution per minute then the hinge between the boards will open at sidereal rate.

example by Ed Danilovicz is shown in Fig. 1.32.[156] While it would appear that these mechanisms, driven by a long threaded rod acting as a worm gear, are rather fragile, proper balancing of the telescope to minimize the forces on the gears appears to work. Sector gears capable of driving the telescope for several hours have been used on a large number of telescopes.

The tangent-arm drive has a cousin in the barn-door drive or Scottish mount as shown in Fig. 4.4. (It is not a Scotch mount. Scotch is a liquor. Things from Scotland are Scottish. I had that impressed on me by a very large angry man named Angus with red hair one evening in Edinburgh.) The device got its name from an article by G.Y. Haig describing a mount made of two boards connected by a sturdy door hinge which are pushed apart by a $\frac{1}{4}$ inch diameter, 20 thread per inch rod turned by a simple 1 rpm clock motor. The 1 rpm motion is easily coupled to small clock motors. The rod must be threaded into the wood 28.97 cm (11.4 in) from the hinge. Other distances for metric threads may be calculated. The hinge pin is aimed at the North Celestial Pole.[157] I constructed one in about 30 minutes and it supports a wide field camera used in searching for synchronous satellites and taking general astrophotos. Like the classical tangent-arm drive, it suffers from drive-rate inaccuracies toward the end of the threaded rod, limiting its usefulness to about a 20 minute exposure.

Several methods have been used to compensate for the tangent error in a Scottish

---

[156] An alternative to casting or cutting worm gear teeth is the hose-clamp drive, using a large steel band hose clamp designed to be tightened using a worm screw acting against partial threads stamped in the metal band. Such hose clamps are commonly found on automobile radiators and they can be purchased in sizes up to half a meter in diameter. See *Telescope Making*, No. 30, Summer, 1987, p. 20.

[157] *Sky & Telescope*, April, 1975, p. 263. See also *Journal of the British Astronomical Association*, Vol.85, No. 5, August, 1975, p. 408.

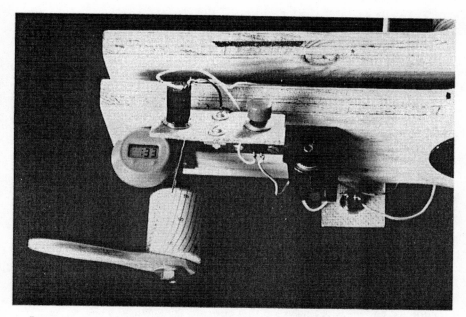

Figure 4.5. Scottish mount compensated for tangent error. Photo courtesy of Frank Zullo.

mount. Frank Zullo's hand-driven adaptation of the Scottish mount attempts to minimize the tangent error with a calibrated drum.[158] His mount, shown in Fig. 4.5, was designed with astrophotography in mind. The observer turns the crank manually, keeping the time indicated on the drum in step with the clock. The time indicated on the drum is skewed toward the end of the exposure in order to compensate for the tangent error. Frank reports that while the system works well, by the time the observer is interested in small errors of this magnitude, there are other, larger errors to worry about such as aligning the mount exactly with respect to the North Celestial Pole, flexure in the mount itself and keeping exactly in step with the clock.[159] A similar empirical correction to the tangent error for a Scottish mount was designed by Gerard Cutting. He used a plastic cam in place of the conventional threaded rod which raises the hinged piece. This was shown at Stellafane.[160]

As an approximation to minimize the errors at the end of the screw travel, one may cock the drive screw at a slight angle so that it is perpendicular to the moving board half-way through the exposure. The maximum error is then half of what it would be if the bolt were perpendicular to the stationary board. Other approximations such as enlarging the end of the screw, and shaping the bottom of the driven

[158] *Sky & Telescope*, October, 1985, p. 392.
[159] An analysis of the allowable errors in astrophotography using a Scottish mount is in an article by Ken S. Hulme in *Telescope Making*, No. 7, Spring, 1980, p. 45.
[160] *Sky & Telescope*, November, 1986, p. 531.

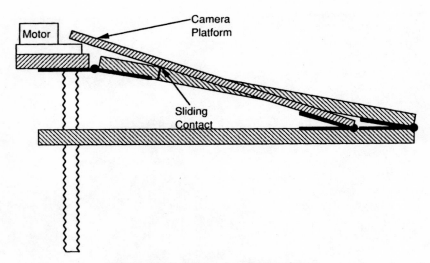

Figure 4.6. Double-arm drive by David Trott.

board to correct for the tangent error have been suggested.[161] Dave Trott has performed an extensive analysis of the errors in the Scottish mount or, as it is sometimes known, the barn-door drive. Trott's result is the double-arm drive, shown in Fig. 4.6.[162] Each of these solutions is an approximation which approaches, but does not match, the tangent-arm error. Even stepping motor drives have been used in an attempt to vary the rate of the driving motor.[163]

The tangent error can, however, be eliminated completely with the use of a curved-bolt drive which is similar in concept to the curved-rod drive described earlier. In such a design a fixed curved-bolt is used and a threaded collar resting on the bottom board is driven by a geared motor, as shown in Fig. 4.7. While the tangent error is eliminated, the periodic gear error is still present. Most of these mounts, however, are used to support wide field cameras whose resolution isn't fine enough to detect the periodic error.

While tangent arms have usually been applied to small, wide field telescopes, the concept has been used on larger telescopes such as the one shown in Fig. 4.8 by Dick Housekeeper. The telescope has the classical door-hinge polar axis. This particular Scottish mount features a motor-driven declination adjustment and a spherical bowling ball mount for coarse pointing. It was seen at the Riverside Telescope Makers Conference in 1987.[164]

---

[161] Private communications with G. Y. Haig.

[162] For a discussion of several iterations of the drive which ultimately led to the double-arm drive, see *Sky & Telescope*, February, 1988, p. 214.   [163] *Sky & Telescope*, July, 1968, p. 80.

[164] This telescope is also described in *Telescope Making*, No. 23, Summer, 1984, p. 29 and No. 27, Spring, 1986, pp. 9–10.

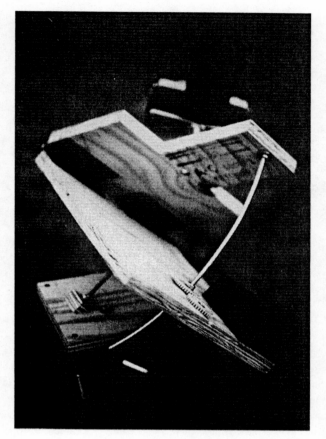

Figure 4.7. Curved-bolt drive.

Manual drives

Remembering that the whole idea of driving a telescope is to produce a fairly fine and controllable motion, there have been some ingenious methods applied to manually driven alt-az telescopes. While such drives generally are not smooth enough for astrophotography, they enable the visual observer to keep the telescope pointed at the object of interest without too much fuss and bother. David Levy of Tucson, Arizona, has rigged a fishing-reel line to the end of his small Dobsonian, as shown in Fig. 4.9. Crazy as it looks, the thing actually works. David is a comet hunter and his observing requirements are that he must sweep long horizontal paths across the sky, then sweep the adjacent horizontal path. His observation technique requires that he sweep along straight swaths of the sky. The drive seems to work well for him. At the

119

Figure 4.8. Dick Housekeeper's Scottish mount (barn-door drive) with bowling ball mount.

time of writing he had discovered eleven comets with various instruments. David says he is spin casting for comets.[165]

The fishing-reel drive is a coarse derivative of the fine-motion controls which have been built into several larger Dobsonian telescopes. Since it is relatively inexpensive to build a half-meter class Dobsonian, larger apertures and the accompanying longer focal lengths have appeared in recent years. The longer focal lengths imply smaller fields of view with the requirement that the telescope have smooth, precise pointing capabilities or the observer will find it difficult to get on small objects and track them manually. Richard Schaffer built fine motion control cams onto his Dobsonian and showed it at the Riverside Telescope Makers Conference in 1981.[166] The cams allow for precise pointing of the large instrument in azimuth and elevation. The cams are engaged after aligning the telescope roughly with respect to the desired object. The short lever arms of the cams push against the base as the observer manipulates the long end, thus providing fine control.

[165] I have suspected that the original purpose behind the design of this telescope was to gain entry into a book on unusual telescopes. There is usually a twinkle in David's eye when he shows the telescope to friends. The joke may have gotten out of hand, however, when he demonstrated the instrument at the 1988 Riverside Telescope Makers Conference and won an award. The telescope is also pictured in *Telescope Making*, No. 33, Summer, 1988, p. 27.

[166] *Sky & Telescope*, August, 1981, p. 122. The cams are described in *Telescope Making*, No. 6, Winter, 1979/1980, p. 4 and No. 12, Summer, 1989, p. 29.

Figure 4.9. Fishing-reel drive by David Levy.

Jose Sasian designed a unique manually operated drive for his unobstructed Newtonian telescope[167], also discussed in the section on two-mirror systems and shown in Fig. 1.27 and Fig. 4.10. His ball and groove drive allows the observer to track an object visually for quite some time before the entire telescope must be reset. While the controls may seem a little obscure at first, they are easily learned and, most important, they are within easy reach of the eyepiece. Jose has optimized the pitch of the threads and the distance from the hinge to the ball and groove to yield a good "feel" to the system. In other words, a reasonable amount of turning on the ball moves the field of view gently. The telescope doesn't jerk around the sky at the slightest touch nor does it require endless turning just to keep up with sidereal motion. This optimization allows the observer to concentrate on the stars instead of the operation of the telescope.

Motive power

Having looked at the methods of coupling electric motors to telescopes, we will now examine some of the more exotic motive powers used. Herschel, Lord Rosse and others in the past had large telescopes controlled by lackeys hauling on ropes. The lackeys also moved the observer's platforms, telescope covers and probably doubled as carriage grooms during the day. Lassell's 1.22 m (48 in) telescope was too large for the clock drives of 1861 so he had a gear train with a flywheel constructed, as shown in Fig. 6.1. The gears were operated by an assistant turning a crank once per second, as measured by a large ticking clock. Lassell said that the cranking could be "continued for hours without being oppressive."[168] Perhaps it was not oppressive for the observer but we have no record of the crank-man's opinion on this.

Sadly, lackeys are hard to come by these days but there will always be free cheap labor in the university research environment where graduate students have been used to move and point a radio telescope.[169] I can personally attest that undergraduates have also been used to move telescopes, clean telescopes, refurbish telescopes, etc. The power obviously comes from the students but as to the motivation; well, sometimes I just don't know where it comes from.

As mentioned earlier, spring-driven and weight-driven mechanical clocks were used extensively in the last century to drive telescopes. With the advent of widespread electric power availability, their use has decreased. In portable telescopes, however, the mechanically powered drives are still used as an alternative to hauling heavy batteries and power converters. Falling weights are a common power source in telescope drives but the methods of regulating the rate of fall vary widely.

Bruce Smith of Salt Lake City, Utah, sets the weight on top of a sand-filled

---

[167] *Telescope Making*, No. 38, Fall, 1989, p. 4.

[168] *The History of the Telescope*, Henry C. King, Dover Publications, 1955, p. 222.

[169] Students at the University of Washington manually pointed their radio telescope, actually an old radar antenna, using ropes and pulleys in scenes reminiscent of the days of Herschel and Lord Rosse. See *Sky & Telescope*, February, 1978, p. 181.

Figure 4.10. Ball and groove drive. Photo by Jose Sasian.

cylinder. A small valve at the bottom lets the sand out at the desired rate, causing the weight to fall slowly. A wire attached to the top of the weight pulls on the right ascension axle wheel, thus turning the telescope, as shown in Fig. 4.11. The drive will run for several hours before requiring resetting. This telescope won an award at the Riverside Telescope Makers Conference in 1979 for the best use of simple materials.[170]

If this method of powering a telescope seems rather obscure, it should be remembered that a similar sand drive on a home-made mount was used by Shelburne Wesley Burnham on his 6 inch Clark refractor while discovering about 400 double

[170] The sand drive is described in *Sky & Telescope*, August, 1979, p. 110.

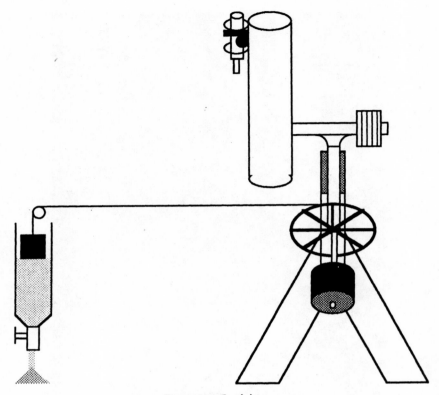

Figure 4.11. Sand drive.

stars in the late 1800s.[171] Henry Draper also used a sand drive in a slightly different configuration. His drive moved a photographic plate holder with respect to a stationary telescope.[172] He also tried a water-clock at one time.

A similar weight-driven principle using hydraulic fluid as a medium rather than sand or water was built by Harold Fisher of Mount Hermon, Kentucky.[173] Hydraulic drives are not entirely new, one having been built by B. A. Stevens in 1955.[174] His drive also used a weight to drive a hydraulic piston. The piston was connected to an arm which pulled a cogged rack across the teeth of the right ascension wheel. As with the other hydraulic drive and the sand drives, the power is not transmitted through the fluid medium. The fluid is used merely to slow and control the rate of the falling weight.

Tom Fangrow invented the hiss drive which works on air. He eliminated the

[171] *Sky & Telescope*, May, 1944, p. 11.
[172] *The History of the Telescope*, Henry C. King, Dover Publications, 1955, p. 269.
[173] *Sky & Telescope*, June, 1978, p. 535.   [174] *Sky & Telescope*, December, 1955, p. 87.

falling weight by making the weight of the telescope press against the fluid medium. A common automobile tire inner-tube supports a polar-aligned hinged platform for the telescope (actually a modified Poncet mount). The air is let out of tube in a controlled manner with a needle valve as shown in Fig. 4.12. The telescope descends slowly, tracking the stars. This drive probably isn't as messy as the sand and hydraulic drives and "recharging" the drive is done with a simple bicycle pump.[175]

A clepsydra (water-clock) drive was used by Donald Menzel and Fernando de Romana for the 1972 solar eclipse in Canada. Their 15.24 cm (6 in) aperture refractor was mounted so that it was front heavy. The objective end rested on a bicycle pump filled with water. A hose clamp regulated the rate at which water left the pump, thus allowing precise tracking.[176] Such makeshift drives and mounts are often seen on solar eclipse expeditions. Indeed, some eclipse telescopes are assembled for only one observation and then never used again – but some parts of them keep reappearing at each appulse. One such telescope was torn down and rebuilt so many times that the owner dubbed it a Phoenix telescope, referring to the mythical bird which repeatedly consumes itself in flames and then rises from its own ashes.

Solar-powered telescopes may seem at first to be a little ridiculous since astronomers generally want to observe when the Sun is down. However, a solar-powered telescope for solar observations was demonstrated by Carl Resick at the Illinois Astrofest in 1983.[177] The drive for his 15 cm (6 in) refractor was powered by a 30 watt solar-cell array. There are also telescopes which are indirectly powered by the Sun. A solar-cell array has been used to recharge the batteries of observers who drive out into the Arizona desert to observe. This allowed members of the Saguaro Astronomy Club to observe for several successive nights during a holiday weekend without having to drive back to town to recharge batteries. There are also several observatory buildings which are heated by solar power.[178]

Miscellaneous drives

Possibly one of the strangest drives of all involves the French telescope array of 1.52 m (60 in) telescopes operated by CERGA (Centre d'Etudes et de Recherches Géodynamiques et Astronomiques). These telescopes were also discussed in chapter 3. Designed by Antoine Labeyrie, each telescope has a spherical base and tube made of concrete. The base sits on a ring of pads which support it. A second ring of pads can be lifted up to support the telescope and is capable of moving the telescope in a sidereal motion for short durations. The second-ring motion is controlled by computer-directed hydraulic pistons with optical encoder feedback. This drive

---

[175] The hiss drive was demonstrated at the 1986 Riverside Telescope Makers Conference and it is described in *Sky & Telescope*, September, 1985, p. 273 and *Telescope Making*, No. 27, Spring, 1986 p. 9.

[176] The bicycle-pump drive is shown in *Sky & Telescope*, April, 1976, p. 252.

[177] *Telescope Making*, No. 22, Spring, 1984, p. 42.

[178] The Lowell Observatory solar-heated observatory is described in *Sky & Telescope*, August, 1976, p. 84.

Figure 4.12. Hiss drive by Tom Fangrow.

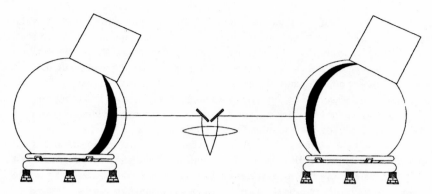

Figure 4.13. CERGA interferometer. The beams from each telescope are combined inside the observatory building, not shown here, to produce interference fringes.

system can track a star for several minutes before it must be reset. The second ring is then lowered so that the telescope rests on fixed pads mounted on the base ring. The pointing of the second ring is then changed before raising it to move the telescope again. The rings of pads can be used to "walk" the telescope to any portion of the sky and then drive it sidereally.

The telescopes are used to measure interference fringes which indicate the size of stellar disks and the separations of very close binary stars. Beams from separate telescopes, shown in Fig. 4.13, are brought out of each instrument and combined in a separate observatory building where sensitive detectors measure the fringes. The angular resolution of the system is equivalent to a single telescope whose aperture is equal to the separation of the telescopes, which can be up to 58 m (190 ft). This corresponds to an angular resolution of about 0.002 arc seconds. Currently there are two of the telescopes operating at the observatory, located in the French Alps.[179] Ultimately, five telescopes will be constructed for the array.

This technique is similar to Michelson's interferometers, shown in Fig. 1.39 and Fig. 1.40, but the optical system has been mounted on two separate telescopes. While the use of separate telescopes may seem simpler than constructing a single large telescope system, the baseline between the two telescopes will rotate with the Earth rather than remain fixed with respect to the star being measured. Thus, a path-length compensation system is required. This is accomplished by moving the combining mirrors and lens along rails on a precision optical table under computer control.

As computers become more and more sophisticated, they are applied to increasingly difficult telescope control problems. We are starting to enter the age of astronomy in space. One might imagine that spaceborne telescopes would not

[179] The interferometer is shown in *Sky & Telescope*, April, 1982, cover photo and p. 334, and May, 1990, p. 494.

require drive mechanisms since they are not attached to a rotating Earth. On the contrary, in space there are forces from the solar wind acting on the spacecraft, magnetic fields producing torques which slew the pointing slightly and the reaction forces which result from moving mechanical parts inside the spacecraft. Unique mechanisms for pointing and maintaining the field of view were designed for the Hubble Space Telescope. At the time of writing, it was launched this morning. The conventional method for orienting spacecraft has been to mount tiny rocket motors around the body of the satellite. These, however, require fuel which must be replaced. They also place a tenuous cloud of exhaust gas in the immediate vicinity of the telescope. Remember that the reason for putting the telescope in space in the first place was to get away from such gasses. Rocket exhaust gasses can also become deposited on the optical surfaces. More exotic pointing and tracking mechanisms are required. They include reaction wheels which are spun up and down to produce countering torques. A second approach is to pass currents through large wire loops which act as magnets interacting with the Earth's magnetic field. More exotic ion and plasma rocket motors are being explored along with techniques of modifying the "sail" area of the spacecraft dynamically to interact with the solar wind.

One exotic proposal includes coating the sides of the spacecraft with liquid crystals which can be made either light or dark to reflect or absorb sunlight and thereby vary the photon pressure on the sides of the spacecraft.[180] This technique may be required in order to position two spacecraft precisely with respect to each other in order to form an optical interferometer.

Recently, proposals have surfaced to construct automated unattended telescopes on the far side of the Moon, where stray light and electromagnetic interference are at a minimum. New drives operating at one revolution per month under one-sixth gravity will need to be designed. Only one conclusion is certain; there will always be telescope-drive engineers.

[180] *Sky & Telescope*, April, 1982, p. 338.

5

Moving the whole telescope

While most professional telescopes are designed to be firmly anchored in concrete, many amateurs do not have the use of a telescope dome and thus they must move their telescopes into and out of the house for each observing session. Some amateurs pack the telescope in the car and drive to an observing site in order to get away from bright city lights. The problems of packing and moving telescopes have been approached in many ways.

Some of the smaller telescopes are made to fold into a compact package in much the same way that opera glasses collapse. Indeed, the word telescope can mean any device that extends and contracts along nested tubes, as in antique naval spy glasses. Folding telescopes discussed here are not the toy telescopes found as prizes in cereal boxes. Some ingenious designs in folding optics have been created. Horace Dall made an early 8.3 cm (3.25 in) reflector with a relay lens which eliminated the need for a tube and also resulted in an erect image for terrestrial viewing.[181] Robert Cox, a well-known telescope designer, has produced a pocket telescope based on Dall's concept.[182]

Moving the telescope short distances

A variety of methods have been developed to "muscle" heavy telescopes from the garage to the backyard or car. The rail-mounted Newtonian shown in Fig. 1.16 has conventional casters and the telescope in Fig. 1.32 is obviously meant to be picked up at the South end and rolled to its storage shed. The Grubb refractor shown in Fig. 7.19 was meant to roll on small rails to its observing position. My own eclipse-chasing telescope which fits in its case for travel has removable casters built into the bottom of the case so that I can push the instrument through airports. It sure beats lifting and carrying a 34 kg (75 lb) trunk around. In order to lift telescopes into vehicles, people have resorted to mounting cranes on trailers.[183]

---

[181] See the article in *Scientific American*, December, 1947, Telescopics Column.

[182] A complete description of the telescope is in *Telescope Making*, No. 21, Winter, 1983, p. 6. See also *Sky & Telescope*, October, 1964, p. 196.

[183] A telescope built by Dwight Bast of Yukon, Oklahoma, required a trailer with a hydraulic crane to erect the mount, as described in *Sky & Telescope*, September, 1984, p. 258.

Figure 5.1. Portable telescope mounted on a lawnmower chassis. Photo courtesy of Dr Clyde Tombaugh. Note the curved secondary mirror holder which minimizes the apparent effects of diffraction spikes.

Some telescopes simply must be moved often. Dr Clyde Tombaugh, the discoverer of Pluto, is retired in New Mexico but he still observes often. His 25.4 cm (10 in) f/5 alt-az mounted reflector, shown in Fig. 5.1, cannot see the whole sky from any given position in his yard due to many trees. Rather than remove the trees, he mounted the telescope on an old lawnmower chassis so that he could move the telescope without much effort. The telescope can now easily view any part of the sky and he still enjoys the shade during the daytime. Dr Tombaugh has developed a reputation for making telescopes out of used materials. Following his lead, I have collected quite a supply of potential telescope parts but my wife keeps telling me we have a garage full of junk.

Trailer-mounted telescopes

Larger telescopes and their support equipment are too cumbersome to pack in the car and unpack after each observing session. Although some astronomers have devoted an entire vehicle to the transport, care and support of the telescope, this is a bit expensive.[184] A less costly alternative is the purchase or construction of a trailer.

[184] When Steve Coe, who built the telescope shown in figure 1, needed a new vehicle he first measured the dimensions of his telescope. He then bought what he refers to as a 17 inch f/5 pickup truck.

Figure 5.2. Trailer-mounted Celestron C-16 telescope used for satellite-tracking. Note the unique clamshell dome.

The portable 40.64 cm (16 in) telescope shown in Fig. 5.2 was operated by the Air Force Avionics Laboratory for tracking Earth satellites at remote locations. It has also been used for eclipse work and for proof-testing prototype television sensors. The trailer-mounted telescope has a four-axis mount which can be mechanically isolated from the chassis of the trailer. Jacks can be installed under the mount to lift it off carrying pads on the trailer chassis. Thus, observers walking on the floor of the trailer will not cause vibrations in the telescope mount. This is the same telescope as shown in Fig. 1.31 before reconfiguration as a bent Cassegrain instrument.

Amateurs have always thirsted after the large apertures used by professionals. Meter-class instruments, however, imply fixed observatories. That wouldn't be a problem except most amateurs live in cities with bright skies. Amateurs with smaller portable telescopes don't mind packing all the gear up and driving into the countryside. With larger instruments, however, the task of setting up becomes more cumbersome than the pleasure of observing. Joe Perry of Las Vegas, Nevada, was faced with the problem of either locating his 0.81 m (32 in) telescope under bright city skies or buying a separate piece of property somewhere in the surrounding desert. He solved the problem by mounting the instrument on a trailer and towing it to observing sites. Note that the polar axis can be adjusted for latitude, as shown in Fig. 5.3. This is a necessity since the trailer may be towed to any latitude on the continent. The adjustment also helps when setting up the system on ground that is not level.

One problem with trailer-mounted telescopes is that trailers must have springs and shock-absorber systems to deal with road bumps. This tends to make the telescope mount wobbly. While this defect can be overcome with chassis jacks and auxiliary supports, one option is to remove the telescope from the trailer before use. This isn't done often since any telescope heavy enough to require its own trailer isn't likely to be easily manhandled. There are exceptions, however, and the heated observing hut built by Bill Volna is one, as shown in Fig. 5.4. The 660 kg (1,460 lb) observatory might have been a problem for several strong lackeys to erect. Since Bill employs no lackeys, he designed a special trailer with arms that slip into fittings in the sides of the observing hut. The arms are moved up and down via a battery-powered hydraulic system built into the trailer. One man can set up the observatory easily. Another view of this telescope is shown in Fig. 6.4. The entire observatory can be leveled quickly with three screws. After shooting a couple of stars to calibrate the position encoders, Bill is ready to observe. The only problem with this set-up is that if you want to stay warm during the entire observing session, you have to ride home on your side in the trailer.

While Bill Volna removes the telescope from the trailer in order to observe, the trailer structure on Steve Dodson's instrument (Fig 5.5) forms an integral part of the mount. The 55.88 cm (22 in) aperture f/7.3 telescope is mounted as an alt-az (Dobsonian) but the mounting sits on an equatorial platform similar to a Poncet mount which can track the sky for up to four hours before being reset.[185] The azimuth ring is 2.13 m (7 ft) in diameter to lend stability to the rather long tube. Although the telescope is tall, considerable engineering and design work went into lowering and nesting the three axes to minimize the possibility of tipping the 410 kg (900 lb) of

[185] The telescope was shown at Stellafane in 1981 and is described in *Telescope Making*, No. 13, Fall, 1981, p. 26, No. 14, Winter, 1981/1982, p. 10, No. 17, Fall, 1982, p. 31 and No. 21, Winter, 1983, p. 50. It is also shown in *Sky & Telescope*, August, 1984, p. 168.

Figure 5.3. Trailer-mounted 0.81 m (32 in) telescope. Photo courtesy of Joe Perry.

moving parts over, especially when the equatorial table is near the end of its travel. The trailer has jacks to raise it off the road wheels and keep the mount steady.

While the telescope is aimed manually by pushing on the tube, the equatorial platform is motor-driven via a double Hamon cam drive. This drive uses a shaped cam to translate the regular linear motion of a driven threaded rod into a constant angular motion about the polar axle. Made from simple materials such as pipe fittings and old automobile parts, the telescope is an example of the statement that engineering is the art of the possible. Steve would like to try astrophotography with the telescope but he finds that most of his time is spent showing others the beauty of the skies. He is currently converting it to a 73.66 cm (29 in) telescope.

133

Figure 5.4. Observing hut being mounted on its transport trailer. Photo courtesy of Bill Volna.

An alternative to trailer-mounted alt-az telescopes is the trailer-mounted equatorial, an example of which is shown in Fig 5.6. Andrew Tomer of Lakeview Terrace, California, solved the problem of precisely aligning the trailer body with North and his latitude by making a four-axis mount. The trailer body does not have to be oriented in any particular direction, a feature which helps in selecting a flat, level observing site into which his car can maneuver the trailer. The trailer has two legs at the rear which are folded down to provide a wider base than the normal wheel spacing of a small trailer. After jacking the trailer up on a three-point suspension the lower two axes, which are an alt-az configuration, are used to align the third axis to the pole. When this is done, the lower two axes are not adjusted for the duration of the observing night.

All four axes are motorized and the upper two axes can be controlled by an on-board computer. The alt-az axes are manually controlled. The right ascension and declination drives each have two stepping motors, one for slewing and one for fine motion, connected through a differential gear. The reason for this is to obtain a large dynamic range of speeds, both fast and slow, from stepping motors which have a limited speed–torque range.

The mount carries two telescopes; a 30.48 cm (12 in) f/11 Cassegrain for visual work and a 20.32 cm (8 in) f/5 Newtonian for photographic work. The entire system,

Figure 5.5. Large trailer-mounted telescope with equatorial table. Photo courtesy of Steve Dodson.

Figure 5.6. Trailer-mounted four-axis telescope. Photo courtesy of Andrew Tomer.

including the computer, requires a 500 watt AC electrical generator to run.[186] The telescope uses the optics from a previous incarnation of a trailer-mounted instrument which was too tall for stable towing, too heavy and had no brakes. Andrew took a lesson from that disaster and designed out all of the undesirable features, proving the old adage that your first telescope teaches you as much about telescopes as it does about the sky. It is your second telescope that gets used regularly.

A distinction should be made between portable telescopes and moveable telescopes. While the little rich-field reflector which you can sling over your shoulder is certainly portable, some trailer-mounted systems can also be considered portable provided they don't require extensive set-up procedures or specially prepared, leveled parking pads. The antenna transporters with specialized tractors and railroad ways which are used to move some radio telescopes are obviously moveable, not portable. There is, of course, a hazy area between them where "excessive set-up procedures" is a highly subjective criterion depending on how willing the astronomer is to fool with the equipment before getting on with the observations. Alas, some of us are tinkerers and some of us are observers.

One moveable but definitely not portable system is a pair of trailer-mounted Pfund type telescopes constructed for infrared interferometry in the ten micrometer wavelength region. Image data from the two telescopes are mixed in much the same

---

[186] The telescope has been displayed at both Stellafane and the Riverside Telescope Makers Conference. It is described in *Sky & Telescope*, October, 1977, p. 330, and *Telescope Making*, No. 5, Fall, 1979, p. 12. It is also featured in *Microcomputer Control of Telescopes*, Mark Trueblood and Russell Genet, Willmann-Bell Press, 1985, p. 127.

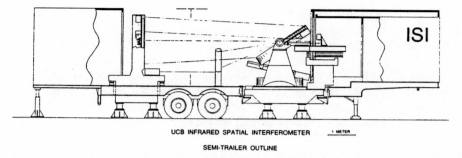

UCB INFRARED SPATIAL INTERFEROMETER   1 METER

SEMI-TRAILER OUTLINE

Figure 5.7. Trailer-mounted Pfund telescope used as an IR interferometer. Illustration courtesy of C. H. Townes and M. Bester.

way that radio telescope data are mixed to interfere and form fringe patterns. A computer analysis of these patterns leads to a high-resolution image of astronomical objects. The angular resolution of the system is about the same as a single telescope with an aperture the size of the separation between the two trailers, although the sensitivity of such an arrangement is related to the actual mirror areas. The entire system taken as a whole is known as the Infrared Spatial Interferometer (ISI). The telescopes are mounted on semi-trailers so that the relative separation and position (azimuth) of the two instruments can be changed easily. Since the telescopes can be transported, they may also be moved at some time to the Southern hemisphere to observe objects which do not rise in Northern skies.

The telescopes, conceived at the University of California at Berkeley, have 2 m (80 in) diameter alt-az mounted flat siderostats feeding fixed 1.65 m (65 in) f/3.14 parabolic mirrors mounted vertically. The beam is then reflected back through a hole in the siderostat to an optical table in the enclosed end of the trailer. The data from each telescope are then mixed or hetrodyned with a carbon dioxide laser before combining them to form the interference image. These Pfund telescopes, named after the German optician who first developed the design, can be lifted slightly off the trailer bed so that they are supported on rigid concrete pads. In order for the system to work as an interferometer, the relative positions of all of the optical elements in both trailers must be controlled to better than the operating wavelength of the telescope. The tolerances on the optical system are thus about one micrometer, requiring a separate visible-light interferometer system to monitor the mount flexure and pointing. The effects of temperature, wind and mount droop are measured by a helium–neon laser system whose rays are shown as the dashed lines in Fig. 5.7.[187] This monitoring laser system is also used to measure the precise pointing position of the telescopes, allowing them to be aimed to an accuracy of about 0.05 arc second.

[187] The telescope is shown in *Sky & Telescope*, December, 1987, p. 576. A more technical description of the system is in *A high precision telescope pointing system*, W.C. Danchi, A. Arthur, R. Fulton, M. Peck, B. Sadoulet, E.C. Sutton, C.H. Townes and R.H. Weitzmann, *Proceedings of the Society of Photo-Optical Instrumentation Engineers, Advanced Technology Optical Telescopes III*, Vol. 628, 1986, p. 422.

At the time of writing, the telescopes were operating at Mount Wilson Observatory at telescope separations of a few meters, although distances of up to a kilometer are theoretically possible. The observations should produce high-resolution images of infrared sources embedded in dust clouds which can be penetrated by the IR radiation. In addition, astronomers hope to resolve detail in bright, compact IR sources such as the center of the Milky Way. The current system angular resolution is 0.25 to 0.03 arc seconds and the theoretical limit is about 0.001 arc second for a 1 km (0.62 mi) trailer separation.

Truck-mounted and vehicle-mounted telescopes

It has often been said that we are an automobile-oriented society. We eat in our cars, we go to movies in them, we often sleep in them. Why not, then, participate in the highest endeavor and observe in them? There is a very logical reason for mounting telescopes on vehicles. Most people live in cities where the background sky brightness inhibits observation of faint stars and nebulae. Some observers pack the telescope in its trunk and load the assembly in a car along with mount, tripod, portable batteries, eyepieces, star charts and other accessories. Then, after reaching the observing site, the whole assemblage must be sorted out and pieced together. Often an hour or more is taken up just getting the telescope ready to observe and this doesn't include driving time or the time required to polar-align the mount.

The 28 cm (11 in) aperture refractor (Fig 5.8) has appeared for many years at the Riverside Telescope Makers Conference. It sat originally on a large pier as shown in the background of Fig. 5.9. The erection of this beast became an annual event at Riverside. Spectators would gather just to see how all of the heavy components were going to be manhandled. With the advent of Comet Halley the owner, Jeff Schroeder, wanted to be able to set up rapidly at public star parties. He remounted the telescope as an alt-az instrument on his car, as shown in Fig. 5.8, and found that he could set up and pack much more rapidly. The mounting borrows features from Dobsonian telescopes such as wooden parts and a Formica® on Teflon® azimuth bearing.[188]

Most vehicle-mounted telescopes such as Schroeder's are designed with the idea that the observer will stand outside the car during observations. Jim Carlisle of San Luis Obispo, California, decided that he would like to sit in the comfortable seat of his car while observing. He mounted the telescope in the Sun roof opening of his car and arranged the eyepiece at a convenient location for observing. The telescope was seen at the 1988 Riverside Telescope Makers Conference. Despite the fact that the vehicle is an Audi, this is not considered a German mount. Note that the large pier-

[188] Mounting details of the telescope are shown in *Telescope Making*, No. 30, Summer, 1987, p. 31. Prior to mounting the telescope on his car, Jeff used to carry the mount parts in a trailer and place the tube on top of the car, as shown in *Sky & Telescope*, March, 1978, p. 257, and *Telescope Making*, No. 4, Summer, 1989 p. 31.

Figure 5.8. Jeff Schroeder's large refractor mounted on a car, as shown at the 1987 Riverside Telescope Makers Conference.

mounted refractor in the background was later installed on a vehicle and also appears in Fig. 5.8.

One of the larger vehicle-mounted telescopes was developed for NASA by Goodyear Aerospace Corporation (now the Loral Corporation). The 61 cm (24 in) aperture f/20 Cassegrain system required a four-axis mount to observe at all latitudes. The lower two axes were manually set to point the third axis at the pole and the fourth axis served as a declination axis. Alternatively, the lower two axes could be left in the stowed position and the upper two axes could drive the telescope in an alt-az mode. The telescope, shown in Fig. 5.10, was mounted on the rear half of the truck bed. The forward half was devoted to an enclosed shelter which housed the controls and recording instruments. When in transit, an observing platform surrounding the telescope folded up and around the instrument to protect it from the elements. The mobile observatory was self sufficient in power, heating and air conditioning. All of the equipment had, of course, been ruggedized in order to withstand the vibration and jolting of road travel.

The system was developed for precision photometry at remote locations. While the original intended purpose was to record and track Earth satellites, the telescope has been used for classical photometry, geodesy and in air-quality monitoring studies.[189]

[189] NASA Contract Report CR-791 and NASA CR-66304, Langley Research Center, Virginia, 1966.

Figure 5.9. Refractor mounted in the Sun roof (Moon roof?) of a car.

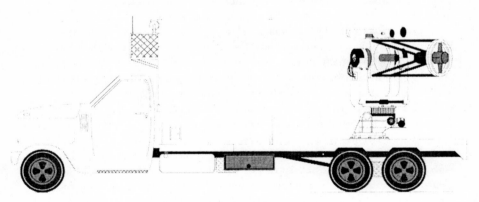

Figure 5.10. Goodyear truck-mounted telescope.

6

The moving eyepiece

In most Newtonians, the eyepiece is located at the sky end of the telescope which isn't bad for a tabletop instrument as Newton originally designed it.[190] Unfortunately, people will tend to take ideas to the limit and the result is an eyepiece which swings all over the observing site, requiring steps, ladders and platforms.

As an example of the complexity required for very large Newtonian observers, the telescope and observing platform by William Lassell, shown in Fig 6.1. approaches an extreme. The telescope was erected on the island of Malta in 1861.[191] The mount is interesting in that it is one of the first to be equipped with a rotating tube assembly, easing access to the eyepiece. The 1.22 m (48 in) f/9.4 speculum mirror performed well and allowed Lassell to make many original observations of faint nebulae, resulting in a catalogue of 600 new nebulae. He and his assistant Albert Marth were able to recognize and document the spiral form of many galaxies. The telescope was also used to search for faint satellites of planets, an extension of Lassell's earlier discoveries of Triton and Hyperion using a smaller telescope.

As with many early large telescopes, there was no dome or shed covering the instrument. Usually the optics and eyepieces were removed during inclement weather. The observer's tower and its doors provided some shelter from the wind but the telescope itself was still susceptible to pointing errors due to gusts. Since the telescope wasn't used photographically, this was probably only a minor inconvenience. The open tube made of iron bars helped with the wind problem but the main reason Lassell used this design was to avoid thermal gradients found in a solid tube. It is an early example of telescope designers paying attention to the finer points of thermal control.

This telescope system is workable only if you have a sufficient number of lackeys to haul on the ropes controlling the turntable and platform motions. The telescope was used for only three or four years before being dismantled and taken back to England. Its history was brief but very productive. At one point there were plans to

---

[190] An excellent description of this pioneering telescope can be found in *The History of the Telescope*, Henry C. King, Dover Publications, 1955, p. 72.

[191] The instrument was originally assembled in 1859–60 near Liverpool in England and later moved to Malta to take advantage of the better observing weather there. See *Sky & Telescope*, October, 1959, p. 664, and *The History of the Telescope*, Henry C. King, Dover Publications, 1955, pp. 220–1.

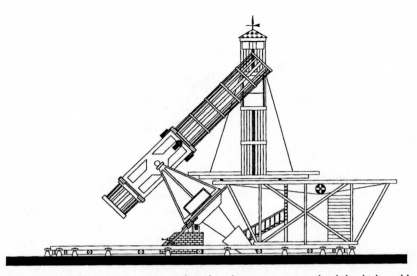

Figure 6.1. Lassell's Telescope. Surrounding the telescope was a circular disk which could rotate like a merry-go-round. A platform was fixed to the disk and the platform held a rolling mount which could be moved toward or away from the telescope. The rolling mount supported a turntable which in turn supported a four-storey observing tower. The astronomer could open the doors of any of the four storeys and insert a plank in any of the notches, providing a convenient place to stand and use the eyepiece. Note the hand-cranked drive.

re-erect the telescope in Australia but the sponsors of the project elected to build an entirely new instrument, the Great Melbourne Reflector, shown in Fig. 3.4.

Now, if you think that telescope is a problem try Herschel's forty foot telescope, shown in Fig. 1.13.[192] This huge machine represented the state of the art in astronomical instrumentation in the late 1700s. Although it had greater light-gathering power than its predecessor, the twenty foot telescope, the forty foot wasn't used as much because of lesser quality optics and the trouble of handling the massive mount. Imagine trying to scamper around the wooden framework in the dark on a cold, damp night when the logs are a bit slippery with dew.

The telescope built by William Parsons, the third Earl of Rosse is perhaps the best example of an unwieldy machine. The 1.83 m (72 in) aperture telescope, shown in Fig. 6.2, was moved by a system of ropes, pulleys and counterweights operated by at least four assistants. The 17 m (56 ft) long telescope tube was mounted between a

---

[192] *Astronomy*, Fred Hoyle, Crescent Books, 1962, p. 166. The remains of this famous telescope are shown in *Sky & Telescope*, March, 1981, p. 194. An engraving and sketch taken just before demolition is in *Sky & Telescope*, March, 1986 p. 253. An excellent description of the telescope written by Herschel is reproduced in *Telescope Making*, No. 3, Spring, 1979, p. 28.

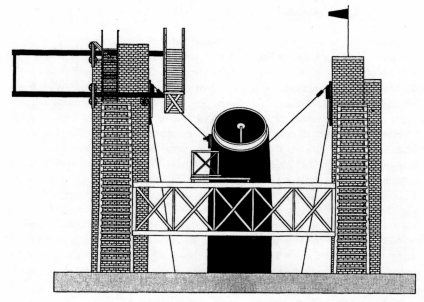

Figure 6.2. The Leviathan of Parsontown, a 1.83 m (72 in) telescope erected by Lord Rosse in the 1840s.

pair of North–South masonry walls and thus it could track stars only when they were within 7.5° of the meridian.[193] A catwalk on the top of the West wall moved East–West to allow the observer to approach the eyepiece. The catwalk was curved in the North–South direction to follow the arc of the eyepiece as it travelled along the meridian. For observations closer to the horizon, an elevator platform was provided at the South end. This elevator rode up inclined rails in front of the telescope.

Since the instrument had no finder telescope, one of the eyepieces was an extremely low power lens 15.2 cm (6 in) in diameter. This yielded a field of view of about half a degree. Some of the earliest experiments in astrophotography were made on the telescope although the slow emulsion speeds available limited work to the Moon and planets. Rosse also designed a large spectroscope and attached it to his telescope for visually exploring the nature of nebulae.

The mirror of speculum metal weighed about three tons. Rosse was faced with technical problems of casting and grinding a mirror larger than any seen before. Several telescopes were first constructed at smaller scales before attempting the 1.8

---

[193] *The History of the Telescope*, Henry C. King, Dover Publications, 1955, pp. 211–7. See also *Astronomy*, Fred Hoyle, Crescent Books, 1962, p. 63, *Sky & Telescope*, June, 1968, p. 366 and November, 1979, p. 489. An excellent description of the telescope and its development is given in *Telescope Making*, No. 7, Spring, 1980, p. 8.

m (6 ft) aperture instrument. He erected a special foundry adjacent to the telescope with three melting ovens and an annealing oven, fired by locally available peat. Workshops were added to make the telescope fixtures and accessories. The mirror required five casting attempts before an acceptable blank was produced. It was then transported on a small railway to the grinding shops and then to the telescope. As with most speculum mirrors, the surface often needed refiguring to remove tarnish. Although the optics remained in the telescope when not in use, the mirror cell was connected with a novel ventilating system to a cache of quicklime, a drying agent.

Rosse collected a small army of machinists, masoners and artisans to support his efforts, which cost him at least twelve thousand pounds. In order to grind and polish the mirror, he designed a steam-powered machine with one of the first temperature-controlled figuring facilities. His estate surrounding Birr Castle became the equivalent of today's Kitt peak with several of the world's larger telescopes close by.

While the machine may seem clumsy by today's standards, it was the first in a line of instruments used to map nebulae and discern spiral structure. It also resolved several nebulae into the individual stars of globular clusters. The telescope operated intermittently up until about 1900 in spite of chronically bad weather and fog from a nearby bog. Although technically an amateur astronomer, Lord Rosse made significant advancements in the state of the art in his day. Astronomy can always find a place for a few more amateurs of his caliber.

Riding the telescope

There comes a time when it's just too much effort to follow the eyepiece around and it makes more sense to attach the observer to the end of the telescope. This has been done in telescopes as small as 30.5 cm (12 in) in aperture. One example is Pierre Schwaar's merry-go-round telescope, shown in Fig. 6.3.[194] The implementation of the telescope is interesting in that the observer's chair remains opposite the eyepiece for all orientations of the alt-az mounted telescope. Note the battery used as a counterweight for the observer. The entire apparatus is moved by electric motors controlled by the observer. The telescope was originally designed to be erected on a portable pedestal but it was later modified so that it was mounted on a trailer, facilitating mobile observations. Having observed with this telescope, which has an excellent optical system, I would say that the only item missing is a calliope.[195]

Bill Volna, a resident of the cold Minnesota climate, overcame the problems of shivering while observing by constructing a heated observing hut which rotates with the alt-az mounted 15 cm (6 in) telescope, as shown in Fig. 6.4.[196] Both the right

[194] *Telescope Making*, No. 4, Summer, 1979, p. 36.

[195] *Sky & Telescope*, August, 1979, p. 178.

[196] The telescope was seen at Stellafane in 1988 and is pictured in *Telescope Making*, No. 34, Fall, 1988, p. 33. A more complete description is given in *Telescope Making*, No. 38, Fall, 1989, p. 29. It has also been shown at Astrofest, as described in *Astronomy*, December, 1989, p. 16.

Figure 6.3. Pierre Schwaar's merry-go-round trailer-mounted telescope.

ascension and declination axes are driven by variable-rate electric motors with three-speed clutches. The entire observatory rotates in azimuth while only the optics move in elevation. A synchro-repeater on the azimuth axle indicates to the operator where he is pointing in azimuth. The elevation circle surrounds the eyepiece and is thus right in front of the observer. In order to find a star, Bill enters the right ascension and declination of the star into a pocket computer and calculates the current azimuth and elevation. He then points to the indicated position and looks. The entire system, accurate to within a degree, needs no finder telescope.

For transporting the telescope, Bill backs a specially built trailer up to the observatory, slides two bars into fittings attached to the observatory and then tips the whole assembly neatly onto the trailer. The trailer is also shown in Fig. 5.4. The telescope has been used at temperatures as low as − 27.8 °C (− 18 °F) without dew problems.

Figure 6.4. Heated observing hut with alt-az drives. Photo courtesy of Bill Volna.

7

The stationary eyepiece

This chapter describes the search for a way to make the telescope move, but not the observer. During the last century it was difficult for astronomers to obtain life insurance. They tended to be sickly from standing out in the damp night air and, more importantly, they tended to fall off their telescopes in the dark. Large telescopes were surrounded by ladders, scaffolding and moveable platforms. James Nasmyth was one of the earlier designers to tackle the problem in the 1840s. Cassegrain telescopes were used in his time but, like the large refractors, they required a tall pier to raise the telescope high enough that the astronomer could stand under them just to get to the eyepiece. Nasmyth took an alt-az mounted Cassegrain and added a third mirror just in front of the primary mirror. The third mirror directed the beam out of the side of the tube horizontally through the elevation bearing. The type, originally called a Cassegrain–Newtonian, eliminated all eyepiece motion when the telescope was moved in altitude. The observer still had to walk around the scope as it was moved in azimuth but this was a great improvement over earlier ladders, ropes, pulleys and scaffolding.[197] In some larger Nasmyth telescopes a seat for the observer is added to the azimuth platform so that the astronomer can ride the telescope.[198]

Truly fixed eyepiece telescopes generally require one or more mirrors to fold the light path through the declination and right ascension axles, as is seen in John Gregory's telescope in Fig. 1.46. Such instruments, also discussed in the section on coudé telescopes, have the advantage of ease of use but the many optical components can each absorb a little starlight and refractive relay lenses can introduce aberrations.

Putting the observer in the center

Several attempts have been made to allow the telescope to swing around the observer, who remains at the center of rotation. This has been done for both alt-az

---

[197] *The History of the Telescope*, Henry C. King, Dover Publications, 1955, p. 219. See also *Astronomy*, Fred Hoyle, Crescent Books, 1962, p. 57.

[198] An example of a Nasmyth telescope with a rider is shown in *Sky & Telescope*, June, 1963, p. 331. This is a telescope which tracks space satellites.

Figure 7.1. Typical Nasmyth telescope.

and equatorially mounted instruments. Zeiss produced a comet-seeker refractor telescope of 20.32 cm (8 in) aperture with an elaborately counterweighted equatorial mount which placed the observer's eye at the center of rotation of both axes. The thought behind this was to make the observer as comfortable as possible, since comet hunting is tedious work.[199] The result was a complicated set of axles and counterweights.

The Great Treptow Refractor 68 cm (26.8 in) aperture f/21 telescope at Archenhold Observatory in East Berlin, shown in Fig. 7.2, is my favorite unusual telescope. It is a massive example of "iron and rivet" engineering, weighing 120 metric tons with counterweights the size of small automobiles. The equatorially mounted telescope was designed to be used by the public as an educational instrument. For this reason, the tube was mounted so that its eyepiece sits at the center of rotation of both axes. The observer stands at center of rotation and the

---

[199] The telescope is described in *The Telescope*, Louis Bell, Dover Publications, 1981 edition, pp. 118 and 120.

Figure 7.2. The Great Treptow Refractor. Photograph courtesy of Archenhold Observatory.

telescope swings around him. Walkways with safety rails have been constructed for use by people who are not familiar with climbing around large telescopes in the dark. This telescope also has the world's longest refractor tube in existence. The original structure dates from 1896. The telescope was damaged by bombing in World War II.[200] It was restored in 1959 and still operates at Archenhold Observatory for public use.[201]

Turret telescopes

Another approach to putting the observer at the center of things is to attach the telescope to the dome of the building. Turret telescopes were popular in harsher climates such as New England since the observer could stay inside and remain warm. Russell Porter, one of the more prolific telescope designers of this century, was

---

[200] One possible explanation for the heavy bombing of the observatory is that from the air, the telescope looks like a huge anti-aircraft cannon.

[201] A history of Archenhold Observatory including the Great Refractor is detailed in *Sky & Telescope*, July, 1984, p. 5. See also *The History of the Telescope*, Henry C. King, Dover Publications, 1955, pp. 307–8.

Figure 7.3. Hartness House turret telescope exterior. Photo courtesy of Rick Rotramel.

involved in the design and construction of at least two turret telescopes in and around Springfield, Vermont.

The less famous of the two is the Hartness House turret telescope. It was built for James Hartness, an early benefactor of the Springfield Telescope Makers and Stellafane. The 25.4 cm (10 in) aperture refractor, shown in Fig. 7.3 and Fig. 7.4, sits on the lawn of the Hartness House Hotel and was built about 1910.[202] In order to move the telescope in right ascension, the whole turret is rotated. Declination motion is provided by swinging the tube on an axis connected to the turret. Windows in the turret allow the observer to see the rest of the sky and roughly align the telescope, although large setting circles are provided inside the turret. The interior of the telescope, shown in Fig. 7.4, had all of the controls to operate the telescope. It even had a radiator to heat the enclosure and keep the observers warm. Mortar over brickwork provides the rounded shape of the building. The observatory was connected to the main house via an underground tunnel. Restoration of the telescope and hotel have been completed by the Springfield Telescope Makers. Many old and innovative telescopes are on display in the house which has been converted into a hotel.

The more famous of Porter's two turret telescopes is the familiar centerpiece of Stellafane, the meeting grounds of the Springfield Telescope Makers in Springfield, Vermont.[203] Each year, in the late Summer, telescope designers and builders from

---

[202] *Sky & Telescope*, November, 1975, cover and p. 333.
[203] The optical arrangement of the telescope can be seen in *Telescope Making*, No. 5, Fall, 1979, p. 8.

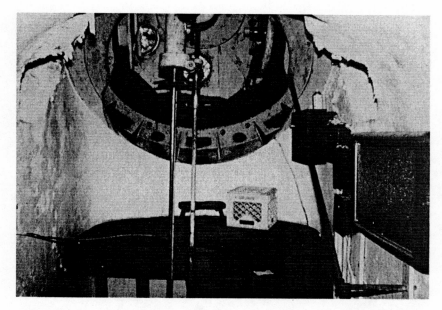

Figure 7.4. Hartness House turret telescope interior. Photo courtesy of Rick Rotramel.

around the world converge on this small town to discuss new designs, show off recently completed instruments and view the stars. Both the Hartness and Stellafane telescopes have a roughly hemispherical dome tilted to 90° minus the latitude of the telescope. Thus the main dome rotation axis points to the North Celestial Pole. The Stellafane telescope on Breezy Hill, however, is a reflector, as shown in Fig. 7.6.[204] Although it was built in 1930, the idea for a reflecting turret telescope occurred to Porter much earlier.[205] A 1920 water-color in Fig. 7.5 demonstrates the basic concept plus the architecture and style which fascinated Porter.

Starlight striking the flat tilted mirror located near the dome is reflected out to the parabolic primary mirror at the end of the struts. The beam is then directed back through the hole in the flat to the eyepiece inside the dome. The flat rotates about an axis connecting the eyepiece with the primary in order to change the declination. The flat is controlled from inside the dome and the whole assembly, including the

[204] A description of Porter's Stellafane turret telescope can be found in *Sky & Telescope*, November, 1954, p. 18, February, 1974, p. 101, October, 1981, p. 371, November, 1984, p. 403 and on the October, 1972 cover. It is also described in *The History of the Telescope*, Henry C. King, Dover Publications, 1955, pp. 427–9. The story behind the development of Porter's turret telescopes is told in his biography, *Russell W. Porter* by Berton C. Willard, Bond Wheelright Publishers, 1976, distributed by Sky Publishing, Inc.

[205] Porter described his ideas for a reflecting turret telescope in *Popular Astronomy*, May, 1921.

Figure 7.5. Proposed reflecting turret telescope. Water-color by Russell W. Porter.

struts and secondary, move together in right ascension. There are several declinations where the struts will intersect the beam but the obscuration and diffraction caused by this shouldn't be any worse than a conventional secondary mirror spider.

When the actual telescope was built in 1930, grandiose architecture and landscaping were scaled down to fit the realities of hard economic times, as shown in Fig. 7.6. In the original design, a second refracting telescope similar to the Hartness instrument was to have been built and attached to the opposite side of the dome from the flat secondary. Thus, two astronomers could have observed simultaneously. The

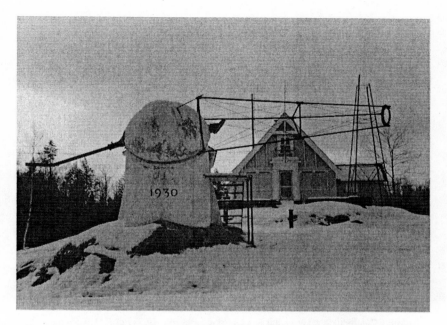

Figure 7.6. Porter's turret telescope at Breezy Hill, Stellafane. The Springfield Telescope Makers' clubhouse is in the background. The pipe mount in front of the clubhouse was used to test early versions of the Springfield mount. The tower at the right once held a siderostat which projected an image into the clubhouse. Photo courtesy of John F. Martin.

refracting telescope was replaced by a pipe and counterweight to balance the turret. This is the first reflecting turret telescope ever built. The interior of the Stellafane turret, shown in Fig. 7.7, has been maintained and it is used during the annual meeting of telescope designers. As with the Hartness House turret, the enclosed observing space keeps astronomers out of the elements. The Breezy Hill turret, however, is not heated in the winter.

The Porter turret telescope on Breezy Hill is a splendid example of the capabilities of a group of amateurs operating in a club environment. Concrete was mixed by hand and the heavy 2 m (6.6 ft) diameter casting for the turret ring was set in place without a crane, as shown in Fig. 7.8. The optics, ground by Porter for an earlier telescope, were remounted by the club members. The telescope is still in operation and is often used.

Russell W. Porter has come to be a symbol for all amateur telescope makers. His innovative designs and dedication to popularizing astronomy with simple, easy-to-build telescopes have inspired many observers to build their own telescopes. He even contributed a humorous generic term to the field of telescope making. Any telescope which is overdesigned, too heavy or just plain cumbersome has come to be

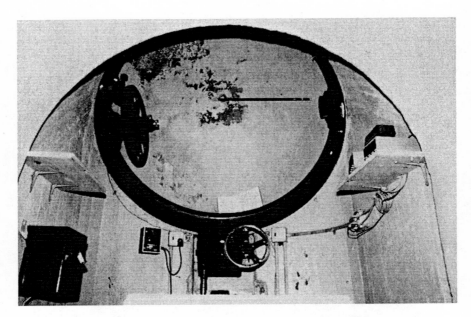

Figure 7.7. Stellafane turret telescope interior. Photo courtesy of Rick Rotramel.

Figure 7.8. Turret telescope at Stellafane under construction. Russell W. Porter, is at the center, wearing a hat. Photo courtesy of Springfield Telescope Makers.

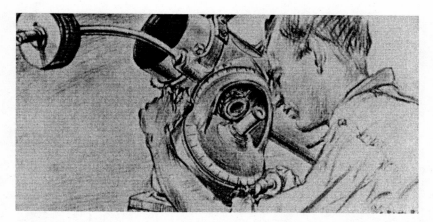

Figure 7.9. The original concept for the Springfield mount. Drawing by Russell W. Porter.

known as a "Porter's Folly telescope".[206] Porter had grand plans to build larger public telescopes with the Springfield Telescope Makers but instead he was called to go off and participate in the design of a slightly different project; the Palomar 5 m (200 in) Telescope. While he visited at Stellafane in later years, his main energies were devoted to the 200 inch telescope.

Springfield mounts

The Springfield mount is basically a Newtonian telescope mounted at the eyepiece end. The reason for this method of mounting is to keep the eyepiece stationary with respect to the observer while the telescope swings around the sky. For this reason, the optical path is brought out through the declination axle, as shown in Fig. 7.9. The first Springfield telescope is reputed to have been designed by Russell W. Porter and built by Oscar Marshall in 1920.[207] The 20.32 cm (8 in) instrument is still owned by the Springfield Telescope Makers of Springfield, Vermont, which also sponsors Stellafane. This instrument can be seen at Hartness House.

In the original Springfield mount, shown in Fig. 7.10, the optical axis isn't quite coincident with the declination axle but the concept of minimizing the motion of the

[206] The original Porter's Folly telescope is described in *Amateur Telescope Making-1* by A. G. Ingalls, p. 135. While fairly conventional in concept, the fork-mount design was made from sand, concrete and scrap iron.

[207] For a description of the development of the Springfield mount, see the biography *Russell W. Porter* by Berton C. Willard, Bond Wheelright Publishers, 1976, distributed by Sky Publishing, Inc. The type is also described in *The History of the Telescope*, Henry C. King, Dover Publications, 1955, p. 428, and *Sky & Telescope*, January, 1947, p. 22 and December, 1954, p. 71. Photos of Marshall using the original Springfield telescope and Porter with a later version appear in *Sky & Telescope*, February, 1974, pp. 102–3, November, 1984, p. 402 and November, 1985, p. 491. Construction of the first Springfield telescope may have started as early as 1915, as indicated in *Sky & Telescope*, October, 1977, p. 336.

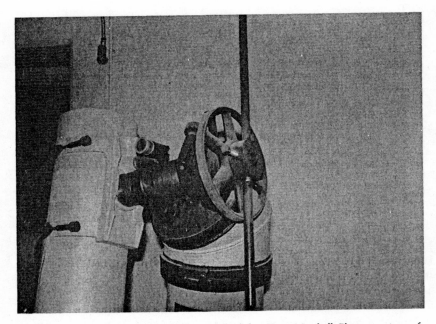

Figure 7.10. The original Springfield mount built by Oscar Marshall. Photo courtesy of John F. Martin, V.

eyepiece was incorporated. Thus, it is not a true Springfield in the strictest sense. In later versions, the optical path was brought out through the declination axle and then bent with a small mirror so that the light traveled up the polar axle, thus assuring that the eyepiece did not move no matter where the telescope pointed in the sky. Although some light is lost due to the extra folding mirrors or prisms, Porter reasoned that the added comfort in viewing would convince more amateurs to take up the pastime of astronomy. Since the observer looks down the polar axle, the viewing position is comfortable and compatible with note taking or sketching. In addition, the setting circles are often mounted adjacent to the eyepiece, making location of objects in the sky easy. It certainly beats having to get up and walk around to the South bearing of a yoke mount to read the hour circle.

On the other hand, the orientation of the sky rotates with respect to the eyepiece as the telescope is slewed and this can be confusing to the novice. Some observers have resorted to elaborate calculations and tables in order to determine which way North and East are in the eyepiece.[208] In a Springfield, it is advantageous to minimize the distance between the right ascension and optical axes. This is because the

[208] An article in *Sky & Telescope*, September, 1981, p. 277 shows how to determine image rotation in Springfield, coudé and other fixed-eyepiece mounts.

light must travel between the two. If the axes are separated widely, a larger portion of the primary mirror's light cone is taken up in the distance between the secondary and tertiary mirrors and thus the secondary mirror must be larger, causing greater obscuration of the incoming light. Since the two axes should be close, the telescope tube has a greater tendency to strike the pier than a similarly sized German equatorial mount.

The only serious problem in the Springfield mount is that the main counterweight is just about at eye level and not exactly in line with the tube. In the dark, it has been known to reach out and whack unsuspecting observers in the head. If you don't believe this is a problem, I have a small scar behind my right ear that I'd like to show you.

The Springfield telescope was designed specifically to be built and used by amateurs. Toward that end, Porter engineered the mount so that it could be built from simple and available materials. For instance, the tube was often made from stove pipes and the pier from water-well casing tubes. The only difficult pieces were the complex right ascension and declination bearings and collar. Porter and the Springfield Telescope Makers made casting patterns of these parts available for use by amateurs. Later telescope makers were able to fashion similar mounts from iron pipe fittings.

Large Springfield mounts with long tubes require the center of rotation to be elevated quite high, lest the mirror end strike the ground. In some of the larger examples, a ladder is required to reach the eyepiece, as seen in Fig. 7.11, or elaborate pier arrangements are required to support the observer.[209] Scrambling around in the dark on ladders and platforms is probably not what Porter had in mind when he designed this telescope whose hallmark was supposed to be ease of use. Thus, there is a practical limit to the size of a classical Springfield before the eyepiece becomes too high for comfortable viewing.

While most Springfield mounts support Newtonians, this is not always the case. The problem of a tall pier required for longer focal length Springfields was solved by Oscar Knab by making his telescope a tri-schiefspiegler Springfield rather than a Newtonian. The instrument, shown in Fig. 7.12, has the optical path exiting the tube in the middle of the telescope, making balancing of the telescope much simpler. This telescope has seen several reincarnations, having later been rebuilt into a wall-mounted telescope with its eyepiece inside Oscar's house and the optics outside in the cold.[210] Oscar generally refers to the schiefspiegler by its German name of brachyt telescope.

---

[209] Elaborate measures such as a moveable chair attached to the pier have been adapted by Clarence Custer Jr for this, see *Sky & Telescope*, May, 1958, cover, May, 1988, p. 556 and November, 1988, p. 464. *Sky & Telescope*, December, 1953, p. 58 shows an arrangement where the observer must hug the mount and stand on pegs sticking out from the pier.

[210] The telescope is written up in *Sky & Telescope*, August, 1958, p. 533.

Figure 7.11. Trailer-mounted 30.48 cm (12 in) Springfield telescope seen at the Riverside Telescope Makers Conference in 1982. Note the ladder required to reach the observer's position. The telescope was built by J. W. Simpson and Russell W. Porter in 1934.[211]

[211] *Sky & Telescope*, August, 1982, p. 185.

Figure 7.12. Oscar Knab's tri-schiefspiegler Springfield telescope. Photo courtesy of
Oscar Knab.

Figure 7.13. Cassegrain Springfield telescope located at the Hyatt Regency Hotel on the island of Maui, Hawaii. Photo courtesy of William C. Leighty.

Another non-Newtonian Springfield mount can be found at the Hyatt Regency Hotel near Lahaina, on the island of Maui, Hawaii. This telescope was conceived by Brent Gordon of La Jolla, California, and William C. Leighty of Juneau, Alaska, as a user-friendly telescope. The 40.64 cm (16 in) f/12 Cassegrain telescope, shown in Fig. 7.13, has a third mirror at the declination axis which directs the light to a fourth mirror which sends it up the polar axle. It is also called a coudé–Cassegrain. This is a robust telescope with 20.54 cm (10 in) bearings on both axes, meant to take the handling typical of instruments used by the public.

The reason for this configuration is to enable easy access to the eyepiece, especially for handicapped observers and people in wheelchairs. This concept is also responsible for placement of the telescope on the tenth floor of a tourist hotel. In order to make viewing easier for the inexperienced observer, a binocular eyepiece is used. While the lenses and prisms of the double-eyepiece system cut down on the total light reaching the eyes, it is felt that the increased ease of use makes up for the loss. Since the main reason for the telescope is to get people interested in astronomy, everything possible has been done to make the telescope easy to use, including a computer control system and a joystick which allows the observer to find objects easily. The computer has a library of about 1000 interesting objects and a sophisticated graphics display system. The telescope is integrated into a program of

Figure 7.14. Tim Parker's wooden Springfield telescope.

celestial tours for hotel guests including a slide show and other instruments such as wide field binoculars. Thus, the telescope functions as part of a teaching and public education system rather than as a scientific research instrument.

The Springfield mount has been seen in many variations and in many materials. One of the finer examples, often seen at the Riverside Telescope Makers Conference, is by Tim Parker of Southern California. The wooden telescope tube, shown in Fig. 7.14, has been crafted as carefully as a fine piece of furniture from Indian rosewood and plywood. The instrument also features a novel placement of the finder scopes inside the cowling which supports the counterweight, giving the telescope a very clean overall appearance. This also places the finder at a convenient position for the observer.[212]

Coudé mounts

Contrary to popular belief, the Coudé telescope was not invented by Jean-Baptiste Christian de La Coudé. The type was introduced by Maurice Loewy of the Paris Observatory in 1891. In French, the word *coudé* actually means elbow. The usual reason for a coudé system is to keep the image plane at a fixed position while the

---

[212] The telescope is also described in *Telescope Making*, No. 12, Summer, 1981, p. 24.

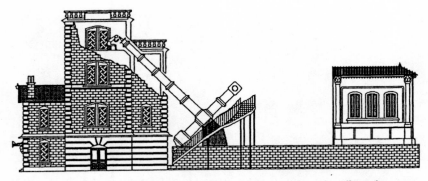

Figure 7.15. Paris Coudé Refractor. The smaller building at the right rolls to the main
building in order to cover the telescope when not in use. Illustration copied from a
drawing courtesy of the Paris Observatory.

telescope slews around the sky. This is accomplished by a set of mirrors and
sometimes prisms which fold the optical path. Some astronomers consider the
Springfield mount to be a sub-category of the coudé. A coudé may be either a
refractor or a reflector. The classic coudé telescope is shown in Fig. 7.15. It is, if
nothing else, the standard unusual telescope. The instrument, located at the Paris
Observatory, allows the observer to remain in a warm room while the telescope is
out in the cold. This is one of two similar instruments, the first being a 26.7 cm (10 in)
built in the late 1800s and the second, shown here, a 59.7 cm (23.5 in) which was built
later.[213] The telescope was used both visually and for photographic work. Two
different objectives were made for these tasks, optimized to either the eye or the film
spectral sensitivity of the day. One might think that observing in close proximity to
a heated building would be detrimental to the seeing. On the other hand, the light
path had to go over many other heat sources in Paris. The telescope was constructed
before it became popular to locate observatories on the tops of mountains. The
photographic objective glass did, however, eventually make it to the top of a
mountain and is installed at Pic-du-Midi Observatory on a different telescope
mount.

A second reason for the coudé design is to avoid dome seeing.[214] The
Kiepenheuer–Institut Für Sonnenphysik constructed a solar observatory on the
island of Capri. The 35 cm aperture f/12.8 three-element apochromat refractor uses a
domeless design, as shown in Fig. 7.16, to avoid seeing problems associated with air
currents in the dome. In order to assure good seeing, the entire optical path is

[213] *The History of the Telescope*, Henry C. King, Dover Publications, 1955, p. 432. The earlier smaller
instrument is described in *Astronomy*, Fred Hoyle, Crescent Books, 1962, p. 59. It is also described in
Splendour of the Heavens, Rev. T.E.R. Phillips and Dr W.H. Steavenson, Robert M. McBride & Co.,
1925, Vol. II, p. 791.   [214] *Sky & Telescope*, November, 1964, p. 272 and May, 1966, pp. 253–7.

Figure 7.16. Domeless coudé refractor for solar work. Photograph courtesy of Kiepenheuer–Institut Für Sonnenphysik. The longer cylinder emerging from the rotating joint is the optical tube. The shorter cylinder is a counterweight.

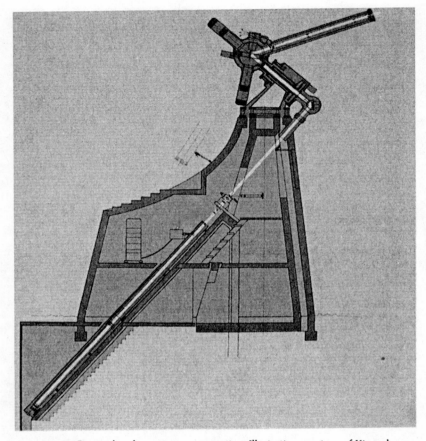

Figure 7.17. Capri solar observatory cross-section. Illustration courtesy of Kiepenheuer–Institut Für Sonnenphysik.

carefully temperature controlled. Plane mirrors direct the beam down the polar axis and then into the observatory building as shown in Fig. 7.17. The main instrument of the telescope is a versatile spectrograph used in studies of the Sun. Since its construction in 1963, seeing conditions on the island of Capri have deteriorated and the Kiepenheuer–Institut Für Sonnenphysik has moved operations to a new telescope on Tenerife in the Canary Islands.

Siderostats, heliostats, coelostats and uranostats

Since telescope designers are an ingenious and somewhat lazy crowd, they often attempt to avoid the work of swinging the whole telescope around the sky. Usually this is accomplished by mounting some arrangement of moveable mirrors in front of the telescope aperture. The main body of the telescope may then be firmly fixed to the Earth. Thus, only the smaller mirror must be moved, resulting in considerable savings in mount costs. Often, the telescope is mounted horizontally pointing North, as shown in Fig. 8.1. The siderostat is a single flat mirror mounted in front of the objective which directs the beam of the telescope to the desired position in the sky. The mirror itself may be either equatorially mounted or driven at varying alt-az rates, in which case it is called a uranostat. Uranostat telescopes do not necessarily have the horizontal telescope pointing North. In the middle latitudes uranostats are used rather than siderostats in order to be able to look at the celestial pole. A uranostat, like any alt-az mounted instrument, has trouble tracking through the zenith. Since the siderostat cannot look at the celestial pole easily, it is most often employed for solar, lunar and planetary studies where the objects of interest stay near the ecliptic. One drawback to using a mirror for pointing is that as the mirror tracks an object in the sky, the image rotates around the optical axis at the focus of the telescope. If you're doing stellar photometry with a single diaphragm then this is not a problem. If, on the other hand, you are making long exposure astrophotos then the star images will look like arcs centered on the guide star.[215]

If the siderostat looks at the Sun then it is called a heliostat but mechanically it's identical to the siderostat. An example of this is shown in Fig. 7.18. The solar telescope, built by Andreas Tarnutzer for a school in Luzern, Switzerland, has an equatorially mounted siderostat and then a folding mirror which directs the beam to a horizontal telescope. The 15 cm (6 in) aperture f/24 objective lens of the telescope is actually between the siderostat and the folding mirror. The light path is enclosed to prevent air currents from disturbing the seeing. The telescope is used for filtered white light observations of the Sun by students.[216]

There is one method of preventing image rotation but it has a limited application. If a polar mounted siderostat mirror is adjusted so that its flat surface is fixed parallel to the polar axis then any adjustment in declination must be accomplished by moving the telescope tube with respect to the mirror. This eliminates one axis of motion of the telescope, the other being taken care of by the rotation of the mirror about the polar axis. In this configuration, called a coelostat, the image does not rotate when the telescope tracks the sky. Owing to the limited applicability of the system, it is used only on solar eclipse expeditions, where the telescope is set up for only one specific observation at one declination before being disassembled.[217]

The polar siderostat was first proposed in 1682 by Boffat of Toulouse as a means of rigidly mounting the extremely long refractor telescope tubes of that time.[218]

---

[215] This problem has been overcome in several large siderostat telescopes by rotating the film holder at the image plane, as in the Great Paris Refractor, described in chapter 8.

[216] The telescope is described in *Sky & Telescope*, July, 1985, p. 71.

[217] A general article on mirror-fed telescopes is in *Sky & Telescope*, January, 1964, p. 46 and July, 1985, p. 73.   [218] *The History of the Telescope*, Henry C. King, Dover Publications, 1955, p. 60.

Figure 7.18. Solar siderostat telescope. Photo courtesy of Andreas Tarnutzer.

Since the achromat hadn't yet been invented, refractors usually had a single lens element of very high f number to minimize chromatic aberration. Thus, the tubes were up to 50 m (164 ft) long and were very unwieldly, as can be seen in Fig. 8.3. In a polar siderostat the telescope tube is used as a polar axle, aligned with the Earth's axis of rotation. In order to see anything other than the pole, a mirror was placed just in front of the lens. The mirror was fixed to the rotating telescope tube so that it tracked stars. A major problem with such telescopes is that the field of view rotates with respect to the telescope tailplate as the telescope is driven, making it difficult to do long exposure photography.

While Boffat's design placed the flat mirror at the upper end, a variation is the polar telescope which has the flat at the bottom end. The example shown in Fig. 7.19 was designed by Sir Howard Grubb about 1880.[219] This 10.16 cm (4 in) aperture telescope is typical of heavy industrial engineering for the period, using castings and rivets for fabrication. The massive assembly moved on rails and could be pushed from inside the observatory part way through a special opening in the wall until the tube was outside in the cold but the eyepiece was still inside out of the weather. The

[219] The telescope is described in *The Telescope*, Louis Bell, Dover Publications, 1981 edition, p.119.

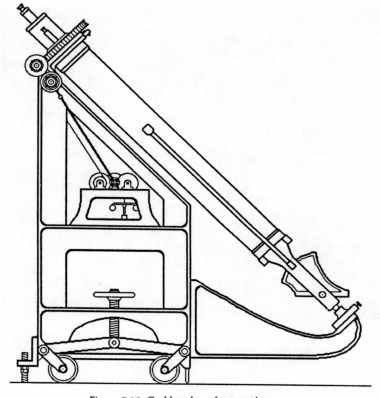

Figure 7.19. Grubb polar refractor telescope.

wheels were then cranked up and the telescope settled onto a prepared set of supports.

The telescope tube, which also acts as the polar axle, looks down through the refractor lens and into a flat mirror mounted in a cradle at the lower end of the assembly. The cradle rocks up and down to change the declination. The entire tube and cradle assembly rotate together in order to vary local hour angle. Only the far Northern declinations are obscured by the telescope tube but the instrument was primarily used for viewing solar system objects which generally stay near the ecliptic.

A variation of the coelostat uses two mirrors, one to track the astronomical object of interest and the other to direct the beam toward the main optics. This is a useful configuration if the telescope and its instrumentation are heavy or cannot be moved. A typical example of this design is the Snow solar telescope at Mount Wilson Observatory, shown in Fig. 7.20 and Fig. 7.21.

Figure 7.20. Snow solar telescope at Mount Wilson Observatory. The 18.3 m (60 ft) and 45.7 m (150 ft) tower telescopes can be seen in the background. Photo courtesy of the Observatories of the Carnegie Institution of Washington.

The telescope, conceived by George Ellery Hale was the largest solar instrument of its day. Its parts were hauled by mule up a precarious trail to the top of Mount Wilson in 1904 before the rest of the observatory was fully developed. The telescope uses a coelostat rather than a polar heliostat so that the image does not rotate while tracking the Sun. The coelostat sends the beam to one of two primary mirrors with focal lengths of 18.3 m (60 ft) and 43.6 m (143 ft). These in turn direct the image to spectrographs or other instrumentation. Throughout many decades of operation, a variety of state-of-the-art solar instrumentation has been developed and used with the telescope. Since the instrumentation does not have to be attached to a moving telescope, it can be massive in size. A key feature of the telescope is the 45.7 m (150 ft) pair of rails set in concrete which forms a gigantic optical bench on which to mount the optics and instrumentation.[220] The rails extend South to the edge of a precipice where the coelostat almost hangs off the edge of the cliff. The reason for this positioning is to place the telescope aperture at a point where the atmospheric

[220] See *The History of the Telescope*, Henry C. King, Dover Publications, 1955, p. 324, and *The Telescope*, Louis Bell, Dover Publications, 1981 edition, p. 126.

Figure 7.21. Snow solar telescope interior at Mount Wilson Observatory. Photo courtesy of the Observatories of the Carnegie Institution of Washington.

turbulence is a minimum. Hale had looked into the problems of air turbulence near the ground and at the edges of mountains.

Many features were designed into the telescope to enhance the seeing. In addition to selecting one of the finer astronomical sites and placing the telescope near the cliff edge, the entire structure surrounding the optics was thermally controlled. While the building was well ventilated to prevent temperature gradients from forming, the optical path was shielded from the effects of the wind. The outer covering was then painted white to prevent the surface from absorbing or radiating heat, a practice still followed in current observatory architecture. Inside the telescope building white drapes were placed at critical points, as shown in Fig. 7.21, to keep air currents to a minimum. Electric fans were used to cool the mirrors which could become heated by sunlight.

The meticulous attention to the engineering details of telescope design resulted in one of the finest solar telescopes which is still in operation today, more than eighty years after its construction. Hale brought the technologies of architecture, thermo-dynamics, weather instrumentation and environmental control to bear on the problem of telescope design optimization. Such a multi-discipline approach was new to astronomy at the time.

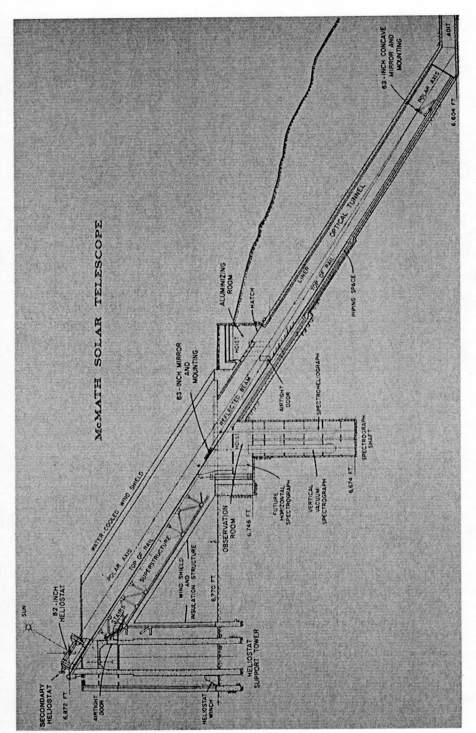

Figure 7.22. McMath solar telescope at Kitt Peak. Illustration by National Optical Astronomy Observatories.

Engineering considerations were beginning to drive the very shape of teles-copes.[221] The cost of large telescopes mandates lengthy studies of theoretical structures, sophisticated computer simulations and the construction of accurate models. Indeed, before construction of the Palomar 5 m (200 in) telescope a one-tenth scale fully functioning engineering model of the instrument was built, including all of the drives and controls. This 0.5 m (20 in) aperture telescope was used for years by students as a teaching instrument.

A direct descendent of the Snow telescope is the McMath solar telescope at Kitt Peak, shown in Fig. 7.22. The same principles which drove Hale's design also figured in the creation of the McMath solar telescope.[222] Like the Snow telescope, it has a long fixed gallery fed by a moving mirror. Unlike the Snow telescope, the long optical path is directed down the polar axis via a heliostat to the primary mirror. Thus one of the mirrors of the Snow's coelostat is eliminated but the image in the McMath rotates as the telescope tracks the Sun. As with other solar telescopes, the beam may be directed to a variety of spectrographs and other instrumentation.

George Ellery Hale was interested in solar observations and was also instrumental in advancing the state of the art in telescope design. He had developed a solar telescope in 1927 with a coelostat mounted on an 18.3 m (60 ft) tower. This instrument is shown in Fig. 7.20 next to the Snow telescope. The picture was taken before a small dome was added to the top of the tower and the structure was modified to minimize vibrations.[223] The idea behind mounting the entrance aperture of the telescope so high was to get the optical path out of the air turbulence near the ground.[224] Armed with success from his new approach, he designed a similar tower 45.7 m (150 ft) high.

One problem with the design was that the wind whistling through the girders of the tower could set up vibrations which would shake the optics at the top. In order to prevent this he designed two towers, one inside the other, as shown in Fig. 7.23. The four main legs, the central pier and all the cross-braces are surrounded by box-like structures which do not touch the telescope supports. When the wind blows at Mount Wilson Observatory, the outer skin may sway several centimeters but the inner skeleton which holds the optics remains steady.

Like the earlier tower telescope, a siderostat directs Sunlight down a central shaft. The 30.5 cm (12 in) aperture triplet lens of 45.7 m (150 ft) focal length feeds a

---

[221] Today, dome models and some telescope structures are even tested in wind tunnels in order to determine their effectiveness, as mentioned in *Sky & Telescope*, September, 1989, p. 251.

[222] The telescope is described in *The Amazing Universe*, Herbert Friedman, National Geographic Society, 1975, p. 60.

[223] See *The History of the Telescope*, Henry C. King, Dover Publications, 1955, p. 338, and *The Telescope*, Louis Bell, Dover Publications, 1981 edition, p. 128. An illustration of the tower after modifications is in *Sky & Telescope*, March, 1985, p. 197.

[224] A popular story concerning the development of tower telescopes reports that Hale explored the concept by climbing a tree with a portable instrument to test seeing conditions far from the ground. See *Sky & Telescope*, May, 1983, p. 453.

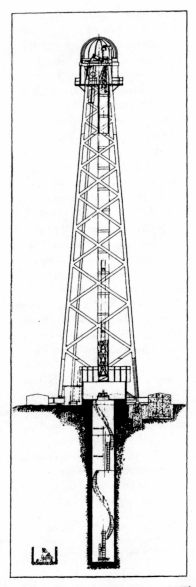

Figure 7.23. The 45.7 m (150 ft) tower telescope at Mount Wilson Observatory. This telescope also appears in the background of Fig. 7.20. Illustration courtesy of the Observatories of the Carnegie Institution of Washington.

spectrograph and spectroheliograph. All of the glass surfaces are surrounded by thermal baffles or are actively water cooled to maintain optical figure. The spectrograph is mounted in a deep pit under the tower which helps maintain a constant temperature for the instrument.[225] It is a characteristic of many large solar telescopes that at least part of the instrumentation is placed underground to minimize thermal problems.

In a combination of the precise thermal control of the McMath and Snow telescopes with the concept of elevating the entrance aperture in order to minimize air turbulence, the next logical step is to eliminate air from within the telescope. The vacuum solar telescope at Kitt Peak, shown in Fig. 7.24, directs the beam through a vacuum tank and into a vertical spectrograph. Owing to the large vertical extent of the telescope, it would be easy for heating and cooling air currents to form inside the building. Since the beam travels in a vacuum, it cannot be distorted by such currents. The telescope is used daily to make magnetograms of the solar disk.[226]

Telescopes built into houses

Unlike the turret telescope in which a dome carrying the optics is built around the observer, in this approach the telescope is built into an existing structure. As with the turret approach, this allows the observer to get inside out of the weather. While this usually refers to the cold night air, in my case I occasionally want to get out of the hot night air since Summer temperatures in Tempe, Arizona, can still be above 38 °C (100 °F) at midnight. Since the telescope must, by definition, be adjacent to a building, there is always the problem of air currents caused by the building being warmer or cooler than the ambient air. While some builders of professional observatories have taken elaborate measures to assure that the outer surface of the building is the same temperature as the air, this is not usually feasible when installing a telescope in a home or classroom building.[227] The usual telescope designer can, at best, merely insulate the building well.

Oscar Knab, a designer of several unusual telescopes, built a wall-mounted 15.24 cm (6 in) f/18 refractor into his home in Indiana, as shown in Fig. 7.25. The

[225] A description of both the 18.3 m (60 ft) tower telescope and the 45.7 m (150 ft) tower telescope is in *The History of the Telescope*, Henry C. King, Dover Publications, 1955, p. 338.

[226] A similar vacuum solar telescope is located at Sacramento Peak Observatory, New Mexico. It is described in *Sky & Telescope*, December, 1969, p. 368. The telescope is unusual in that the entire vacuum tube, spectrographs and instruments rotate in azimuth to follow the Sun and are floated on mercury bearings.

[227] The Air Force Avionics Laboratory's Cloudcroft Observatory uses a double-walled dome building where ambient air is drawn in between the outer skin and the inner structural wall. This air is then forced down an underground tunnel and exhausted at some distance from the observatory. Thus, the outer skin of the building is at ambient temperature. The building is so well insulated that even during the winter, the air conditioners must be run just to get rid of the heat from the machinery and people working inside.

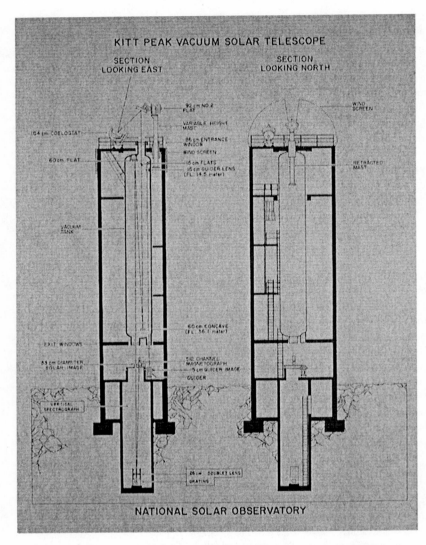

Figure 7.24. Vacuum solar telescope at Kitt Peak. Illustration by National Optical Astronomy Observatories.

Figure 7.25. Wall-mounted refractor telescope. Photo courtesy of Oscar Knab.

telescope, which could be classed as a polar refractor, is fed by a 25.4 cm (10 in) flat siderostat, whose position is controlled remotely from inside the house. Potentiometers on the siderostat axles provide an indication of the mirror position to within 0.5°.[228]

Dealing with the temperature differential from one end of the telescope to the other has several possible remedies. In one version of his telescope, Oscar made the eyepiece holder from wood so that it did not transmit heat from the house to the telescope. The majority of the optical path is outside the heated space so the telescope would essentially be at ambient outside temperature. Only the last few centimeters of the light path are in warmer air. Since the refractor is a closed-tube design, there are minimal problems with air currents in the tube. While the top of the telescope tube is in the warm house and the bottom is in the cold air, Oscar theorizes that the warm air floats on the cold air and thus prevents mixing air currents in the tube. It would be interesting to see if in the Summer, a cooler house and warmer exterior would result in tube currents. The telescope has been in use for several years, protecting its owner from the winds of Winter and the mosquitoes of Summer.

While a polar telescope with a siderostat mounted to a building suffers from an image rotation and a limited swing in declination, the advantages in comfort often outweigh the difficulties.[229] If the primary interest of the observer is in solar, lunar or planetary work then the inability to see far from the ecliptic is no problem. The image rotation problem can be solved by rotating the entire tube in right ascension. In that way, position angles of double stars can be read with respect to the rotating tube. Similarly, a photographic plate holder attached to the rotating tube will not experience image rotation problems during long exposures. That was the approach taken in the Harvard College Observatory polar telescope, designed by W. P. Gerrish, shown in Fig. 7.26. The 30.48 cm (12 in) aperture instrument with a 45.7 cm (18 in) siderostat was erected in 1900. It was used for years in visual work on variable stars and as a teaching instrument.[230]

An interesting observatory was built into the peak of an existing house by Edward A. Halbach of Estes Park, Colorado. A hole was cut into both the North and South slopes of the roof, as shown in Fig. 7.27. A concrete block isolated pier for the

---

[228] The telescope is described in *Sky & Telescope*, July, 1970, p. 46 and July, 1985, p. 74. It was originally constructed as a schiefspiegler as seen in *Sky & Telescope*, May, 1963, p. 292, and later converted into a refractor.

[229] The declination swing of a polar mounted refractor built into a house is limited in the direction of the pole by either the roof of the building or the proximity of the telescope objective. In the direction away from the pole, a larger declination swing implies that a larger siderostat flat is required to prevent vignetting of the light path. At some declination far from the pole, the cost of the flat mirror exceeds the interest of the astronomer in astronomical objects which may be low on the horizon most of the night anyway.

[230] The telescope is described in *The History of the Telescope*, Henry C. King, Dover Publications, 1955, p. 431, *The Telescope*, Louis Bell, Dover Publications, 1981 edition, p. 122, and *Sky & Telescope*, January, 1964, p. 46.

Figure 7.26. Gerrish student telescope mounted in the side of a classroom building. Photo from Harvard College Observatory.

Halbach Observatory

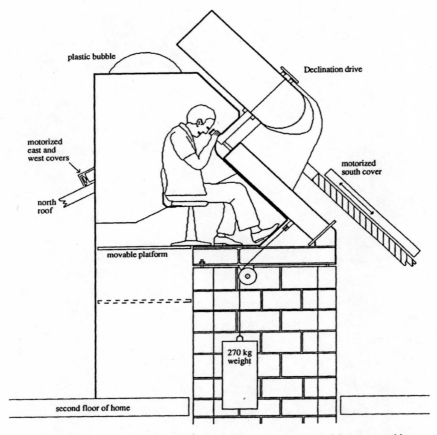

plastic bubble

Declination drive

motorized
east and
west covers

motorized
south cover

north
roof

movable platform

270 kg
weight

second floor of home

Figure 7.27. Roof-mounted insulated observatory. Drawing copied from original by
Edward A. Halbach.

telescope was then built up through the house to support the telescope. The 40.6 cm
(16 in) aperture off-axis Cassegrain telescope was then assembled as a modified
German Mount. The telescope has third and fourth mirrors which bring the focus to
a fixed eyepiece, almost like a coudé telescope. The eyepiece is coaxial with the polar
axis. The declination arm is long enough for the telescope tube to clear an insulated
cupola which protects the observer from the cold. Electric drives and light emitting
diode (LED) position readouts complete the system.

The hole in the roof allows the declination arm and tube to swing to any point in the sky. In order to stow the telescope, the RA is driven Westward and then the tube is pointed slightly down so that the roof covers can be closed. When the East, West and South covers are in place the telescope is sealed away from the elements but the upper part of the cupola protrudes above the roof line.[231]

In my opinion, the ultimate stationary eyepiece involves the use of television sensors. While the professionals have been using TV for decades, the technology is now filtering down to the amateur community. Instead of standing out in the cold trying not to breathe on chilled eyepieces, I sit inside, kick my shoes off, sip a warm cup of cocoa and observe in a much more civilized manner.

[231] A description of the telescope can be found in *Telescope Making*, No. 38, Fall, 1989, p. 40.

8

Limits

Giant telescopes and small telescopes – we have always been interested in the
biggest, the fastest, the tallest and the smallest. Why else are circus sideshows
popular? Note that while writing this chapter I had to revise it twice because new
telescopes came on-line. With the advent of multiple telescope systems and radio
telescopes, the argument over which telescope is largest becomes a bit fuzzy.[232]
While the Russian 6 m (236 in) telescope may be the largest single-element optical
mirror in the world, there are larger radio telescopes. The surface of some of the
submillimeter wavelength reflectors rivals optical quality. Arrays of telescopes
provide larger equivalent apertures in terms of resolution but they are made up of
many very small telescopes. The largest of the arrays is no longer the National Radio
Astronomy Observatory's Very Large Array (VLA) in New Mexico. Construction
has started on their Very Long Baseline Array (VLBA) composed of several large
dishes scattered across the North American continent and reaching out to the
Hawaiian islands. By far the most extensive array of astronomical detectors is
probably the gamma-ray detectors aboard the Vela satellites and some interplane-
tary spacecraft orbiting both the Earth and Sun. These satellites have detectors on
board which can detect the output from gamma-ray burster stars. By comparing the
time of arrival of a burst at several satellites, astronomers can determine the rough
direction to the source. It's not a telescope in the sense that an image is formed but in
that it can detect an event at a distance and roughly indicate the direction to the star.

Large aperture refractors

The Yerkes Observatory 1.02 m (40 in) is the largest refractor telescope, right? We all
know that. Wrong. There is at least one existing telescope (the Great Treptow
Refractor) with a longer tube and there have been at least two larger aperture
refractor telescopes. Although they no longer survive as telescopes since their tubes
and mounts were scrapped, their optics still exist. The Yerkes remains only the
largest aperture *working* refractor.

[232] It can be argued that for years the largest single-mirror telescope in the world was not the Palomar 5 m
(200 in) or the Russian 6 m (236 in) telescope but rather Grote Reber's 9.6 m (31.4 ft) radio telescope
built at Wheaton, Illinois, in the early 1940s. Today there are several radio telescopes in the 100 m
class and one of 305 m (1000 ft) at Arecibo, Puerto Rico.

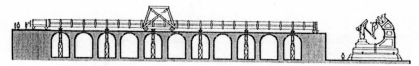

Figure 8.1. The Great Paris Refractor of 1900. The siderostat on the right feeds the 1.25 m (49.2 in) objective at the right end of the horizontal tube.

The second largest refractor ever made was the 1.04 m (41 in) Pulkovo Observatory (Russia) refractor whose mount rusted away before the optics were installed.[233] The unfinished optics are now in the Science Museum in Newcastle upon Tyne, England.

While there are several Schmidt corrector plates of 1.22 m (48 in) aperture[234], the largest true refractor lens belongs to the Great Paris Refractor of 1900. In order to save the cost of making a huge equatorial mount, it was designed as a siderostat-type telescope with a fixed horizontal tube as shown in Fig. 8.1. The instrument was made for the Paris Universal Exhibition, a trade and cultural fair. It therefore operated near searchlights on the fairgrounds with all manner of sky brightness and seeing (atmospheric stability) problems. In the defense of the decision to place the telescope in Paris, note this was before it was well understood that mountain tops provided the best seeing. In addition, the long steel tube without provision for ventilation or thermal control probably induced seeing effects within the instrument.

A separate flat 2 m (79 in) aperture siderostat mirror tracked the stars. The massive siderostat mirror support and its moving parts floated in a pan of mercury in order to assure smooth motion of the assembly. Light travelled through the lens and down a 60 m (187 ft) tube to the focus. Two objectives were made, one for visual work and one for photographic work. The eyepiece holder, shown in Fig. 8.2, and plate holder were mounted on small carriages which rolled on rails in order to focus. From diagrams and photographs of the telescope, it appears that the plate holder was designed to rotate in its mounting around the horizontal optical axis. This motion was to compensate for image rotation induced by the Foucault siderostat as the telescope followed the stars. For visual work, the lowest power available was 500 × . Since the pointing controls were located at the siderostat end of the telescope, a telephone was installed to allow the observer to tell the operator how to move the telescope.

The telescope was used for about one year, during the Paris Universal Exhibition, but it was unsuccessful as a serious astronomical instrument since few scientific observations were made with it. The public paid a few centimes each to look through

[233] See Sky & Telescope, December, 1975, p. 370 and March, 1984, p. 227 for a description of the instrument including the fabrication of the optics. In The History of the Telescope, Henry C. King, Dover Publications, 1955, p. 432 the name of the intended observatory is given as Nikolaieff Observatory in Southern Russia.

[234] The largest cemented doublet is the Schmidt corrector plate of the 1.2 m (47.2 in) United Kingdom Schmidt, see Sky & Telescope, December, 1981, p. 537 and March, 1984, p. 228.

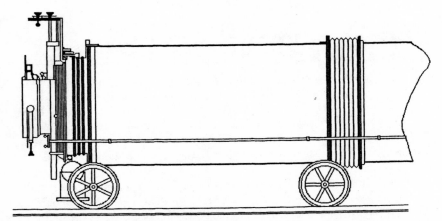

Figure 8.2. Eyepiece detail of the Great Paris Refractor of 1900. The carriage, which could accommodate either an eyepiece or photographic plate holder, was focused by rolling the assembly back and forth on rails.

the telescope but this did not cover the operating costs of the huge instrument. The optical parts are stored at the Paris Observatory and the rest was broken up for scrap.[235]

Long focal length refractors

In the late seventeenth century, color-corrected doublet refractor lenses were unknown and large mirror telescopes had not been developed. In order to compensate for the aberrations of low f number objective lenses, astronomers developed longer and longer focal length instruments with singlet lenses. Eventually, f numbers far in excess of 100 were used. The limit (or perhaps past the practical limit) was a 64 m (150 ft) focal length instrument, shown in Fig. 8.3, erected at Sternenburg Observatory by Hevelius.[236] He also made several other telescopes of this general design with focal lengths of 18 m (60 ft) to 21 m (70 ft). The whole assembly was supported with a mast 27.4 m (90 ft) high.

Several approaches were tried to keep the tube assembly stiff and light. Rolled

---

[235] The telescope is described in *The History of the Telescope*, Henry C. King, Dover Publications, 1955, p. 433, and *The Astronomical Scrapbook*, Joseph Ashbrook, Sky Publishing Corporation, 1984, p. 179. See also *Sky & Telescope*, August, 1958, p. 509, January, 1964, p. 48 and July, 1985, p. 75.

[236] A description of this instrument can be found in *The History of the Telescope*, Henry C. King, Dover Publications, 1955, p. 53, and *The Telescope*, Louis Bell, Dover Publications, 1981 edition, p. 3. There is mention in *The Picture History of Astronomy*, Patrick Moore, Grosset & Dunlap, 1964, of a 64 m (210 ft) telescope used by Huygens and a 182 m (600 ft) instrument designed by Adrien Auzout which apparently was never constructed. 64 m (150 ft) instruments were also used at the Paris Observatory, see *Sky & Telescope*, February, 1980, p. 100.

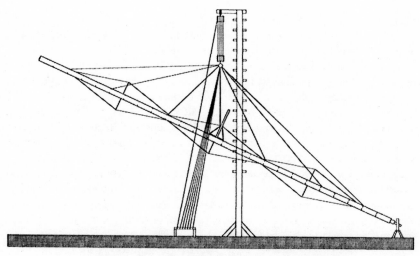

Figure 8.3. Long focal length refractor built by Hevelius.

paper, a common tube material of the day, couldn't be made stiff enough for the giant instrument and a metal tube would have been too expensive and heavy. One approach used two boards, attached along their lengths to form a trough with an L cross-section. This worked well enough on dark nights but during twilight observations or on nights with strong Moonlight, stray light entering the eyepiece decreased contrast. Note that in a refractor of such high f number, contrast isn't excellent to begin with. One of the later solutions was to place a series of black aperture stops along an open-frame tube assembly. From the eyepiece, no portion of the sky could be seen except through the objective. This appears to be the final solution applied by Hevelius.

The addition of the aperture stops imposed the requirement that the tube be maintained straight and not sag in the middle, thus requiring a series of ropes to each section of the tube structure. Eventually, half a dozen ropes supporting the tube were added, each individually controlled by an assistant. If the elevation of the tube changed appreciably during an observation, the tube-support ropes would have to be readjusted for the new tube angle. If I were Hevelius I'd try to make most of my observations near the meridian where elevation changes are the smallest. The telescope must have been a joy to operate in the slightest breeze, swaying to and fro. Its location, near the Baltic sea at Danzig (Gdansk), Poland, doesn't seem conducive to calm weather. The mast, ropes and spars may also have resonated at certain wind speeds like the rigging of a sailing ship.

Tracking, of course, was accomplished manually. The long focal length implied that the eyepiece would have to move about 1.5 mm each second just to keep up with the stars. The telescope hung from its support ropes and had no axles. Thus, it falls in the category of an infinite axis mount. In some of the later drawings of the instrument, it appears that a screw-driven azimuth/elevation table was placed under the eyepiece end to control the pointing of the telescope.

While Halley called the telescope "useless", Hevelius stated that the telescope was easy to use. Perhaps Hevelius had more patience with cantankerous equipment, for he certainly had the enthusiasm to pursue astronomy under adversity. At the age of sixty-eight he rebuilt his entire observatory after it was destroyed by fire in 1679. The telescope was used to make serious observations and Hevelius even made a workable star catalog using the instrument.[237] A similar telescope used about the same time is shown in Fig. 2.14.

The Great Treptow Refractor 68.07 cm (26.8 in) aperture telescope at Archenhold Observatory in East Berlin was mentioned in chapter 7 and shown in Fig. 7.2. The 18.3 m (60 ft) tube is longer than the Yerkes Observatory 1.02 m (40 in) telescope tube but it is not as long as the Great Paris Refractor of 1900. The Treptow telescope remains the longest working refractor.

Large telescope platforms

Many amateur astronomers initially built or purchased a Dobsonian telescope in order to maximize the aperture per dollar spent. Later they became interested in astrophotography but the alt-az mount would not track the stars smoothly unless they moved the telescope to the North Pole. There is a cheaper and warmer solution. The equatorial platforms developed by Adrien Poncet and others have allowed observers to mount large and often bulky telescopes on an equatorial drive. The drive must be reset every hour or so but that is a small price to pay for a cheap and reliable equatorial mount. Not only telescopes but cameras have been mounted on such platforms.

While this practice has been limited mostly to amateurs, there have been some professional equatorial platform installations. The Russians have been tracking artificial Earth satellites for some time using a wide field camera mounted on a "jack leg" platform. The platform is supported by three legs and one of these has a motor which can make the leg longer or shorter. The other two legs have hinges, one at the top and the other at the bottom, arranged so that the hinges line up with the Earth's axis.

One of the larger camera platforms in the world is a modified Poncet type built by Professor Russ Nidey at Yavapai College, Arizona. The platform, shown in Fig. 8.4, is 2.9 m (9.5 ft) on a side and is designed to hold 50 photographic cameras at once. It was built for the recent return of Comet Halley and the idea was that many amateur astrophotographers could make guided exposures simultaneously. The mount,

---

[237] In modern times a reincarnation of the long-focus refractor hung from a pole with ropes has been used. Richard Berry made a replica of the type with a 6 cm (2.4 in) aperture and a 4.06 m (160 in) focal length, see *Sky & Telescope*, February, 1976, p. 130.

Figure 8.4. Large platform for astrophotography. Photo courtesy of Professor Russ Nidey.

which is built from old boiler parts from a nearby smelter, has its polar axle parallel to the Earth's axis. The platform is manually driven by turning a crank. It now functions as part of a yearly astrophotography festival.

Adrien Poncet of France has proposed building a 15 m (50 ft) diameter platform capable of carrying large telescopes or entire star parties. The design is unique in that the underside of the platform would have a large air tank which would be partially immersed in a pool of water. The float would carry about 90% of the weight of the platform and thus ease the burden on the bearings and pivot of the platform.[238]

Multiple mirrors and lenses

I was tempted not to show the Fred Lawrence Whipple Multiple Mirror Telescope (MMT). It has almost ceased to be unusual because everybody's making them these days. The MMT might have belonged on the front cover of this book if it had been written ten years ago.[239] The telescope's basic design concept has been copied to the extent that it is now mundane. The MMT designers have a right to feel just a bit

[238] This platform is mentioned in a long article on Poncet designs in *Sky & Telescope*, March, 1980, p. 257.
[239] In his introduction to the 1981 edition of Bell's *The Telescope*, Dr Jay Pasachoff states that the MMT is "The strangest current design. . .".

proud, for imitation is the sincerest form of flattery. There are, however, many different kinds of multiple mirror telescopes in addition to the famous one on Mount Hopkins. Many multiple mirror telescopes predate the MMT, although they are often functionally different. In this section we will first consider multiple mirror telescopes that are actually multiple telescopes on a single mount. Then we will look at "big dumb light-buckets" composed of multiple mirrors, followed by true coherent imaging multiple mirror telescopes and finally telescopes composed of an array of separate optical systems whose images are combined.

Multiple optical systems on a single mount

Most telescopes have more than one optical system on the mount if you count the finder scope. There have, however, been some in which the smaller optical system is more than just a little telescope used to get the main one on the right object. The most obvious example is the photographic guide scope and many larger telescopes have these. Often the smaller aperture guide scope actually has a longer focal length and hence more magnifying power than the main telescope. Photographic guide telescopes can also have elaborate mounting schemes which allow them to be precisely misaligned with respect to the main optics so that the observer can track a bright star which is not in the field of view of the main telescope.

My own telescope, used for video occultation work, requires a double-finder system because the field of view of the main optics is only a few arc minutes wide. The normal 6° field of view finder telescope is used to slew the object of interest to within half a degree of the desired position. Then I switch to a second finder which is a 90 mm (3.5 in) aperture f/11 Maksutov telescope used at a magnification of about 50 × . With the second finder I can place the object of interest precisely in the 4 arc minute wide TV field of view.

There have been several telescopes which operate in more than one color of light simultaneously. In order to save time when doing multicolor photometry, both a three-channel and a six-channel photometer have been constructed using identical primary objectives for each channel.[240] While this multiplies the cost of the telescope by the number of channels, it saves the time required to measure each star several times while switching different filters in and out of the beam.

The most common example of multiple optical systems on a single mount is the double astrograph. There are several of these at professional observatories around the world. The prime use of such instruments is to take simultaneous astrophotos of the same region of the sky in two different colors. One of the main telescopes would use a film or plate with a blue-sensitive emulsion while the other would use a red-

---

[240] The three-channel photometer with 7.62 cm (3 in) Maksutov optics is described in *Sky & Telescope*, August, 1977, p. 101. The six-channel photometer is described in *Sky & Telescope*, September, 1980, p. 200.

Figure 8.5. Bosscha Observatory, Indonesia, twin astrographs in a single tube. The third aperture is a guide telescope. Photo courtesy of Bambang Hidayat, Director, Bosscha Observatory.

sensitive emulsion. This is useful when studying variable stars which may change in brightness from night to night.

Bosscha Observatory on the island of Java in Indonesia operates a twin Zeiss double refractor, shown in Fig. 8.5. This is an unusual instrument in that it mounts two 60 cm (23.6 in) f/17.9 telescopes and a 30.5 cm (12 in) guiding telescope in a single tube. The telescope was made before World War I.[241] It is used extensively for double-star observations both photographically and visually with a micrometer. Owing to the high humidity of the area, it is equipped with an unusually large dewcap which can be lengthened to as much as 3 m (9.8 ft), in which case it extends beyond the dome slit.[242]

Noncoherent multiple mirror light-buckets

One of the main reasons for placing multiple optical elements on a single mount is to make one larger optical system. Since the cost of large mirrors and lenses goes up

[241] *The History of the Telescope*, Henry C. King, Dover Publications, 1955, p. 347.
[242] *Sky & Telescope*, August, 1971, p. 61 and cover.

approximately with the square of the aperture area, mounting many smaller primaries or objectives should increase the cost linearly with area. This relationship is often overridden by the costs of mounting and collimating many separate optical systems. As a result, there is only a narrow range of applications in which multiple mirrors or lenses is practical. These telescopes are usually designed to perform only one or two tasks well and they are not good general purpose optical instruments. It should be noted that while the sensitivity or light-grasp of a telescope goes up with the number of optical elements added to the primary aperture, the resolution of the overall telescope is a function of the size of each element. In other words, the resolution of a 100 m aperture telescope composed of 1 m segments will be the same as a 1 m telescope.

One successful example of multiple lenses is the lunar ranging telescope (LURE) operated by the University of Hawaii on Mount Haleakala, Hawaii. The telescope is composed of eighty objective lenses, each 19.5 cm (7.675 in) in aperture, as shown in Fig. 8.6. The combined aperture of the lenses is equivalent to a single 1.74 m (68.5 in) aperture telescope. There is a complex arrangement of beam-folding mirrors in order to bring all of the images to a common focus. The telescope is used in conjunction with a separate short-pulse laser transmitter which reflects light beams off the Moon in order to measure accurately the distance to the Moon.[243] The light is bounced off optical retro-reflectors placed on the Moon during manned and unmanned explorations there. Data from the observations are used in a wide variety of applications from basic timekeeping to checking on Einstein's theory of relativity.

Since it is important that the distance from the laser to the Moon and back to the multiple lens telescope be the same for all lenses, path-length compensators are built into the combining optics. This telescope, like several others having multiple optical systems operating in parallel, has been referred to as a fly's eye telescope. The LURE Observatory has also been used to measure the distance to space satellites with reflectors on them in order to measure the size and shape of the Earth and continental drift.

A distinction is made here between multiple mirror light-buckets and imaging systems. Several telescopes of low angular resolution have been constructed where the parameter of prime importance is aperture. Typically these systems are used for wide field photometry or monitoring portions of the sky. They are often used with photoelectric sensors such as photomultipliers. Apertures of 10 m are often possible with such telescopes. True imaging multiple mirror telescopes will be covered in a later section.

An emerging area in high energy astrophysics is the study of gamma-rays and cosmic-rays (charged particles). Gamma-rays with energies of 10^{12} to 10^{16} eV (electron volts) can be caused by supernova explosions or exploding primordial

[243] The subject of lunar ranging is covered in *Sky & Telescope*, October, 1980, p. 277. The LURE telescope and several other laser-ranging telescopes are pictured in the article.

Figure 8.6. Multiple lens lunar-ranging telescope. Photo courtesy of University of Hawaii.

black holes, if they exist.[244] When these energetic particles or waves impact the atmosphere they give off a very fast pulse of light called Cerenkov radiation, usually about 5 nanoseconds in duration.[245] It is the speed of the pulse that allows astronomers to differentiate between Cerenkov radiation and slower transient phenomena such as small meteors and fluctuations in the airglow. Typically, a gamma-ray will cause a volume of air about 30 m (100 ft) in diameter and 10 km (6.2 mi) long to glow. Cosmic-rays and charged particles produce a broader glow about five times larger. Thus, by the size of the glow, the two can be differentiated. The size can be determined by looking at the time duration of the light pulse as it hits the detector. Smaller sources emit much shorter pulses. For gamma-ray sources such as supernovas and exploding black holes, a burst of many individual Cerenkov pulses would be seen over a time ranging from a thousandth of a second to several seconds. There is no sure way to differentiate between a supernova event and an exploding

[244] A description of the search for exploding black holes using large telescopes and light-collectors is given in *Sky & Telescope*, February, 1978, p. 113. A description of the physics behind the experiment is given in *A Brief History of Time*, Stephen W. Hawking, Bantam Books, 1988, p. 111.

[245] Cerenkov radiation is the blue glow produced when a particle or wave passes through a transparent medium such as air at a speed greater than the speed of light in that medium but not greater than the speed of light in a vacuum.

Figure 8.7. 10 m (32.8 ft) optical reflector used for gamma-ray and cosmic-ray studies at the Fred Lawrence Whipple Observatory, Mount Hopkins, Arizona. Photo courtesy of the Smithsonian Institution.

black hole but in order for a supernova to trigger the detector, it would have to be relatively close, say within our own galaxy. We know that, on the average, a supernova this close occurs only once every thirty years. The observers were looking for black hole events which might occur once or twice a day.

A giant light collector was erected at Mount Hopkins in 1968 to study the faint pulses of light produced by these effects. The telescope, shown in Fig. 8.7, is composed of 248 small hexagonal mirrors, each with a focal length of 7.3 m.[246] Each of the small mirrors has an adjustable mount to bring the focus of the entire array to within a 6.35 cm (2.5 in) spot size. While a spot size of 6.35 cm is considered horrible for a classical astronomical telescope, remember that the purpose of this instrument is simply to collect photons, not make high resolution images. The entire array is 10 m (32.8 ft) in diameter. While this is one of the larger telescopes in existence, it also has the shortest range. Even though the particles and waves it detects are from far beyond the solar system, the telescope itself actually looks only at the upper reaches of our own atmosphere. Stars, planets, comets and moons are mere background noise

[246] A description of the instrument is given in *Sky & Telescope*, November, 1968, pp. 281–4. Cosmic-ray research on the instrument is described in the issue of June, 1985, p. 498.

Figure 8.8. Solar Thermal Test Facility which was used as an astronomical telescope. Note the steerable parabolic collectors in the background. Photo courtesy of Sandia National Laboratories.

to it. Note that while this is a multiple mirror telescope on Mount Hopkins, it is not the famous Fred Lawrence Whipple Multiple Mirror Telescope which was installed at a later time.

In order to differentiate between the background of gamma-rays and an exploding black hole, observations are often made simultaneously with a solar collector at White Sands Missile Range, New Mexico, 405 km (250 mi) distant. The pair of detectors is operated as a long baseline coincidence detector. Single random background gamma-rays will cause Cerenkov radiation from only a small part of the sky. An exploding black hole would produce a shower of gamma-rays which would be visible above both telescopes at the same instant. In other words, if both detectors see an event at the same time, it's real, but if only one sees the pulse, it's probably spurious. Thus far, exploding black holes have not been detected but the lack of positive observations can be used to set an upper limit on the rate of primordial black hole explosions.

There are two 11 m (36 ft) parabolic arrays at Sandia National Laboratories in Albuquerque, New Mexico. The light collectors, originally designed to be solar power collectors, are composed of 220 separate spherical-figure mirrors mounted on a parabolic frame. They can be seen in the background of Fig. 8.8. Like the Mount

Hopkins 10 m light-collector, the Sandia collectors are supported on a motor-driven mount. These telescopes have also been used to look for Cerenkov radiation. The pair has been used together as a coincidence detector.

One of the largest optical instruments in the world isn't a telescope. It has been used, however, as a telescope for experimental purposes. The Solar Thermal Test Facility operated by Sandia National Laboratories in Albuquerque, New Mexico, was designed to test various concepts for generating power from the Sun. The centerpiece of the facility is a Solar Power Tower surrounded by 222 multiple mirror heliostats on separate alt-az mounts, as shown in Fig. 8.8. Each heliostat has 25 square mirrors, each 1.22 m (4 ft) on a side. The mirrors are made as flat sheets but the mountings deform them into shallow concave mirrors with focal lengths of 55 m (180 ft) to 207 m (680 ft) depending on the distance from the mirror to the central power tower. The whole array contains 8250 m² (about two acres) of glass and is unquestionably the largest total aperture visual wavelength telescope – so far. The heliostats are each separately pointed by a computer in the control building.[247]

As an experiment in night-time utilization of the facility, several observations were conducted. The first and most obvious was to point the array at a star and use it as a giant light-bucket to measure star brightness. As a photometer, however, there are some technical problems. The first is that the mirrors do not have the same optical quality which one would find in common astronomical telescopes. They were designed to make an image of the Sun, which is half a degree across. Finer detail would entail excessive costs in the manufacturing of the optics. Consequently, a point source image at the top of the 61 m (200 ft) tall tower is about 2 m (6.6 ft) in diameter. A second problem is that the mirrors are supported on arms which sag slightly as the elevation of the arms changes. The sag is less than that required to move the image of the Sun more than its own diameter but it is appreciable. Finally, the arrays of mirrors can catch the wind and move slightly. As with the problem of sag, pointing errors caused by the wind are not great enough to hinder solar power collection but they are sufficient to hinder stellar photometry. The star Vega was imaged onto the tower and a photograph, shown in Fig. 8.9, was taken of the mirror array. Since the image of Vega at the focal plane is large and the camera lens was much smaller, the camera intercepted only a tiny fraction of the starlight directed to the tower. Any photometric detector would need an aperture of 2 m at the focal plane in order to collect all of the light. It would also have to be shielded from the lights of the nearby city.

In spite of the difficulties of large image size and pointing stability, there is a class of observation which requires extremely large apertures. One example is the study of faint Cerenkov radiation caused by either gamma-rays or cosmic-rays (charged particles) hitting the upper atmosphere, as mentioned earlier. The Solar Power Tower was used to look for gamma-rays as evidence of exploding black holes, as

---

[247] The Solar Thermal Test Facility is described in *Sky & Telescope*, April, 1978, p. 287. Astronomical observations are described in *Sky & Telescope*, February, 1980, p. 119.

Figure 8.9. Solar Power Tower mirrors illuminated by Vega. Photo courtesy of Sandia National Laboratories.

previously described. Experiments at the solar facility continue using other large multi-facet parabolic reflector arrays. A significant disadvantage in using the Solar Power Tower and its array of siderostats is that the light path length from a patch of sky illuminated by Cerenkov radiation to the detector is a function of the part of the siderostat array that it encounters. Rays striking the farther mirrors will arrive tens of nanoseconds after rays striking the nearer ones. This makes the discrimination between cosmic-rays and gamma-rays difficult. Parabolic collectors have the advantage that all rays travel the same distance from the source to the detector.

While most "light-buckets" do not have good angular resolution, a telescope can be built with good resolution at only one point in the image plane. Classical telescopes oriented toward visual or photographic work require aberration-free images over a specified field of view. A stellar spectrograph, however, requires high resolution only for the star of interest. Any image around the one star is thrown away. If the telescope is to be used only for stellar spectroscopy then design criteria can be relaxed with a considerable saving in costs. One shortcut is to make the primary mirror a spherical rather than a paraboloidal surface. A second saving can be obtained if the primary is made from many small segments. Such a telescope has been proposed by Pennsylvania State University and the University of Texas.[248] The

[248] The telescope is described in *Sky & Telescope*, February, 1988, p. 129, *Telescope Making*, No. 39, Winter, 1989/1990, p. 6, and *Science*, Vol. 239, February 19, 1988, p. 868.

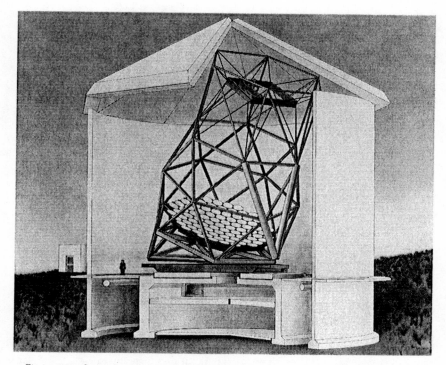

Figure 8.10. Spectroscopic Survey Telescope. Illustration courtesy of the University of Texas McDonald Observatory.

Spectroscopic Survey Telescope (SST) will have 85 very thin mirrors, each 1 m (39.4 in) in aperture, as shown in Fig. 8.10. This is equivalent in light gathering ability to a single mirror with an aperture of 8 m (26.2 ft). In order to overcome the problems of spherical aberration, aspheric 30.48 cm (12 in) secondary and 20.32 cm (8 in) tertiary mirrors will be mounted near the focal plane as a modified Gregorian corrector set.

In order to avoid problems associated with keeping all the mirrors aligned as the telescope is moved, the telescope will be permanently tilted to an elevation of 31° and fixed there. It will be moveable in azimuth only. Thus, a swath of sky from declinations of $-6°$ to $+66°$ will be observable. In order to track an object for spectroscopic integrations, the detector will be moved in the image plane to compensate for the motion of the star. The spectrographic detector will be attached to the telescope only by a fiber optic cable and thus the telescope will not have to support a heavy optical system at its focus.

This telescope design illustrates how defining the observing problem carefully can eliminate many of the general purpose measurement capabilities of a classical astronomical telescope which drive costs upward. The telescope is being built for about $6 million, a factor of ten less expensive than a general purpose telescope with the same aperture. This telescope design also supports the maxim that it is the mark of a good engineer to know where he can cut corners but it is the mark of an excellent engineer to know where he must not cut corners.

True multiple mirror telescopes

This is a case where the initial dream, which was to produce a large inexpensive mirror by making it in smaller sections, failed. Several attempts were made but the problem of mirror support structural flexure was overwhelming. As the telescope was pointed to various locations in the sky, the mechanism supporting separate mirrors would bend slightly due to the force of gravity pulling first one way, then another. It wasn't until the advent of computer-designed structures which did not introduce pointing errors that the concept became viable. At the same time other computer systems which could dynamically control the mirror position became feasible. During the long process of development, however, there were some unusual telescopes created.

William Parsons, the third Earl of Rosse, experimented in the late 1800s with two-piece speculum metal mirrors. One mirror had a central 7.6 cm (3.0 in) aperture disk surrounded by a 3.8 cm (1.5 in) wide annular piece. The combination was secured to a common backing plate and ground to a true sphere. No parabolizing correction was made. The central disk was then moved back slightly to bring its focus into the same position as the outer mirror's focus.[249] The idea behind this experiment was to reduce the grinding time required for parabolizing the mirror. While this is not significant for a 15 cm (6 in) aperture mirror, Parsons was developing techniques for making truly large mirrors up to 2 m (6.6 ft) in aperture. Since he did not use this technique for his largest mirrors, it is assumed that the results were disappointing.

After World War II, G. Horn-d'Arturo assembled a zenith telescope at the University of Bologna with an aperture of about 1.8 m (5.9 ft). It was composed of 61 separate 20.3 cm (8 in) hexagonal mirrors. In order to avoid the problems of supporting the mirror as it pointed to various locations in the sky, it was used only as a zenith telescope but it had a field of view of about 1.3°. Long exposure photographic plates were made by moving the plate holder to counter the sidereal motion. This limits exposures to about 6.5 minutes but stars as faint as magnitude 18.5 were recorded. The intention was to make many of the telescopes and place them at 1.3 degree intervals of latitude so the entire sky could be photographed.[250] While useful scientific data were obtained with the telescope, the design was ignored outside of Italy. Perhaps it was an idea whose time had not quite arrived. Several concepts seen

---

[249] *The History of the Telescope*, Henry C. King, Dover Publications, 1955, p. 206.
[250] *Sky & Telescope*, February, 1978, p. 100. See also *The History of the Telescope*, Henry C. King, Dover Publications, 1955, p. 435.

Figure 8.11. Väisälä's multiple mirror telescope. On the left is the mirror cell structure showing the use of three-dimensional triangular structures. On the right is a view down the tube of the telescope. Photographs courtesy of Turku University Observatory.

in this telescope such as moving the detector at the image plane were later adapted to more modern instruments such as the Spectroscopic Survey Telescope.

Yrjö Väisälä, director of Turku University Observatory in Finland, built a multiple mirror telescope in 1949.[251] The telescope had six 32 cm (12.6 in) aperture f/8 mirrors mounted in a rigid cell, as shown in Fig. 8.11. A seventh mirror sat in the center of the array. Since all mirrors were ground as spheres, a field-flattening and correction lens was required near the focus. The telescope was intended to be a model for a larger telescope that was never constructed. Like the Bologna multiple mirror telescope, it was intended to be used as a zenith instrument or at least a fixed instrument in order to eliminate the problem of the dynamics of mount flexure as a function of elevation.

The design disasters of the 1800s through the 1960s were finally brought to fruition in the Fred Lawrence Whipple Multiple Mirror Telescope (MMT). The instrument, shown in Fig. 8.12, is located on Mount Hopkins and operated by the Smithsonian Astrophysical Observatory. In this design by the University of Arizona and the Smithsonian Institution, a rigorous analysis of the structure supporting the mirrors was performed during the 1970s.[252] A key part of the design was to use an alt-az mount. This assured that the force of gravity would act along only one plane with respect to the telescope framework.

Consider any equatorially mounted telescope scanning the horizon. With the telescope pointed North, place a mark on the top of the tube. Then scan the telescope

---

[251] *Sky & Telescope*, February, 1978, p. 101. See also *The History of the Telescope*, Henry C. King, Dover Publications, 1955, p. 435.
[252] The MMT is described in *Sky & Telescope*, July, 1976, p. 14 and July, 1977, p. 9.

Figure 8.12. Fred Lawrence Whipple Multiple Mirror Telescope located on Mount Hopkins, Arizona. Photo courtesy of MMT Observatory.

Eastward until it points due East. The mark on the top of the tube will rotate until it is on the North (or South for some types of mounts) side of the tube. Continue scanning around the horizon and note that the tube rolls and the mark may, for some azimuth, wind up on the bottom. In order for the telescope optics to remain in place, the mirror cell must be able to handle side forces in any direction. Now consider an alt-az mounted telescope. As it scans the horizon, a mark on the top of the tube will stay on the top of the tube. This mirror cell must handle side forces in only one direction.

The design concept of the MMT was to let the structure bend slightly as the telescope was moved in elevation but to control the amount of bending and compensate for it, as in the Serrurier truss.[253] In addition, active optics were employed in which each mirror's pointing could be adjusted via computer with respect to the other mirrors and the telescope structure. Thus, as the telescope

[253] Classical Serrurier-truss tube designs are shown in figures 1.25, 1.41 and 2.9. The design allows the front structure and the rear structure to sag but the deflection of both parts is calculated to be equal, thus keeping the primary and secondary mirrors aligned.

framework was pointed to different elevations, each of the mirrors was tweaked repeatedly to keep it aligned. Typically, a new alignment is required every few minutes or if the telescope is slewed to a different part of the sky.

The telescope, completed in 1978, has been used for photometry and imagery but its prime use has been in spectroscopy. The design was optimized for infrared work but the telescope has also been used during the daytime as a submillimeter wavelength radio telescope. The instrument has even been used for speckle interferometry of space satellites. The six separate 1.8 m (72 in) f/2.7 mirrors have a combined area equivalent to one 4.5 m (176 in) telescope. It is, effectively, the third largest optical telescope in the world in terms of light-collecting area. The short 4.93 m (194 in) effective focal length contributes to the telescope's compact size. The six images are superimposed at the image plane to form one image. Since a common focus is used, each of the six mirrors had to be fabricated with exactly the same focal length. There are times, however, when the images are not stacked but rather placed in a line, when the light is directed to the slit of a spectrograph.

Moving the images around with respect to each other at the combined image plane can be accomplished with a versatile set of six remotely controlled articulated combining mirrors and associated optics. The beam position from each of the six primary mirrors can be adjusted with respect to the other five to within an accuracy of about 0.06 arc seconds. With the images superimposed, an optical wedge in each beam can then be adjusted to match the total path length from each primary mirror to the optical plane. Thus, the phase information contained in the six optical systems is preserved so that interferometry can be performed. Preservation of the phase information was originally stated as a desirable goal in the design of the instrument. The engineers were not sure, however, that they could measure and compensate for gravity deflections and thermal expansions of such a large and complex structure to within a micrometer. Recently this goal has been accomplished, resulting in enhanced capabilities for the telescope.

One key element of the MMT is the observatory building which shelters the instrument. The entire 500 ton building rotates in azimuth with the mount, although the outer structure is mechanically isolated from the optics. This is to prevent vibrations caused by people walking in the building from interfering with observations. The alt-az mount design and the tight fit between the building and telescope result in a fairly small observatory structure for such a large telescope. The aerodynamic shape of the building and its placement on the summit of Mount Hopkins were carefully considered in the design. The idea was to cause a minimum of turbulence which might degrade the seeing. In addition, the interior of the chamber containing the telescope is thermally controlled and isolated from the spaces housing the control room, offices and conference room. This carefully integrated design considering all aspects of telescope construction resulted in a large, capable instrument for about a third of the cost of a comparable single mirror equatorially mounted telescope.

Ironically, current plans for the world's first successful large multiple mirror telescope include replacing the six mirrors with a single thin 6.5 m aperture spin-cast mirror.[254] The mount and rotating observatory building along with much of the sophisticated instrumentation will be retained. That would make it the largest optical telescope in the world — until some of the 8 m mirrors start coming out of the spin-casting ovens. There are even plans to make multiple mirror telescopes with each optical element an 8 m or 10 m mirror.

At the time of writing there are no less than six multiple mirror telescopes under construction or in use with effective apertures of more than 1 m. All of them use some form of computer-controlled adjustment of each mirror segment. A typical example is the Russian AST-1200 with seven 0.4 m hexagonal segments at the Crimean Astrophysical Observatory.[255] The telescope uses active optics to maintain mirror alignment.

A second and more ambitious project is the Keck 10 m telescope currently under construction on Mauna Kea, Hawaii. It will have 36 hexagonal segments each 1.8 m (72 in) across. As with other multiple mirror telescopes, a computer-controlled active optic design is being incorporated.[256] This approach, which competes heavily with the thin spin-cast mirror approach, appears to be the method by which larger and larger telescopes may be constructed.[257] Segmented mirrors will also be used in space based telescopes which have apertures larger than the current Space Shuttle cargo bay dimensions. Large telescopes will then be assembled after reaching orbit.[258]

Even amateur telescope designers have joined the trend of using multiple mirrors. A fork-mounted telescope using four 15.24 cm (6 in) mirrors and one 20.32 cm (8 in) mirror was shown at Stellafane in 1979.[259] Light from the five mirrors is combined at a single image plane. The total light-gathering power is the same as that from a 36.58 cm (14.4 in) f/7.4 single mirror telescope. As with other multiple mirror telescopes, the folded and combined optical paths make the whole instrument

---

[254] See *Sky & Telescope*, August, 1988, p. 129. Such a mirror would probably require active optics control in order to maintain its figure.

[255] The AST-1200 is described in an article in *Sky & Telescope*, June, 1980, p. 469 dealing with several new telescope designs.

[256] Construction of the Keck 10 m telescope is shown in *Sky & Telescope*, December, 1987, p. 575 and February, 1989, p. 127. The original design is detailed in the March, 1985 issue, p. 223.

[257] The construction of a multiple mirror telescope has even been the subject of a science fiction short story. The space-based telescope was designed to have an aperture of 8 km and use millions of mirror segments. See *The Big Dish* by John Berryman, *Analog Science Fiction/Science Fact*, November, 1986, p. 12. In an even more fantastic prediction of the future, the editors of Time-Life Books have envisioned a telescope made by self replicating nanomachines which transform an asteroid into a parabolic reflector two million miles in diameter. See *Spacefarers*, Time-Life Books, Alexandria, Virginia, 1989, pg 118.   [258] Private conversation, Dr Alan N. Bunner, NASA HQ.

[259] The telescope, designed by Harald Robinson and Ken Laune, is pictured in *Sky & Telescope*, October, 1979, p. 303. It is also described in *Telescope Making*, No. 5, Fall, 1979, p. 11 and No. 11, Spring, 1981, p. 6.

much shorter than a conventional telescope. A total of 19 reflecting surfaces are required to combine the images from all five primary mirrors. The entire assembly is held together in a girder and truss structure which is actually a junked space satellite frame.

Arrays of telescopes on separate mounts

The United States Air Force once tried an experiment to measure the size and shape of Earth satellites by watching them occult or pass in front of stars. A line of several 20.32 cm (8 in) Schmidt-Cassegrain telescopes, each with a separate high speed photomultiplier, was used to monitor bright stars. The stars were selected based on predicted paths of Earth satellites. The probability of at least one satellite of interest occulting a star at some time during a given night is fairly high. This is not surprising considering the thousands of satellites in orbit. The total time that the satellite would spend in front of the star, blocking its light, is very small due to the high speeds of satellites. In practice, it became difficult to predict the exact positions of satellites ahead of time in order to know which star to monitor. This is because satellites are subject to the vagaries of the solar wind which can shift their orbits by several kilometers in a day.

In order to monitor cosmic-rays with energies on the order of 10^{11} to 10^{19} eV you need either a very large detector or a great amount of patience since their arrival rate is about 10^{-16} particles per second per square meter of aperture per steradian of field of view, or roughly one event every 316 million years when viewed with a conventional telescope. Such particles, however, impact air molecules in the upper atmosphere and break them up into a shower of nucleons, mesons, electrons and other secondary particles. These particles in turn excite nitrogen molecules which glow by luminescence briefly. This faint flash can be detected many kilometers away if the observer looks at the correct portion of the spectrum. If an observatory is constructed which looks at the entire sky at the nitrogen wavelength then the detector encompasses a volume of hundreds of cubic kilometers and the probability of seeing an event approaches one or two per month.

The Laboratory of Nuclear Studies at Cornell University constructed the all-sky monitoring system dubbed the Cornell Cosmic House in the late 1960s. The observatory, shown in Fig. 8.13 and Fig. 8.14, stared at the sky surrounding the instrument. No attempt was made to track the stars. The observatory building is a truncated icosahedron. Each of its 16 upper sides has a 45.72 cm (18 in) fresnel lens which images a portion of the sky onto an array of photomultiplier tubes. A typical lens is shown in Fig. 8.13, imaging the observatory. In front of each lens is a filter which allows only the 3100–4100Å light from the nitrogen glow to reach the detectors. Behind each filter is an array of several photomultiplier tubes, each looking at a separate small segment of the sky. The tubes each had special pulse discrimination circuitry so that only fast transient signals were recorded, thus

Figure 8.13. Cornell Cosmic House. Photo courtesy of Dr Edith Cassel.

eliminating brighter but slower moving sources. The entire observatory used 505 photomultipliers whose output was then directed to an array of 505 cathode ray tubes inside the observatory which were then photographed by a single 70 mm film camera.[260]

The observatory could operate any night when the Moon was not in the sky. Results were mixed at the Cosmic House, largely due to the poor observing weather in upstate New York. The demonstration that the technology was in hand to make this type of observation convinced astronomers that they should pursue the project elsewhere. Several of the Cornell researchers continued the project by moving to the relatively clear skies of Utah.

In order to detect cosmic-rays, the University of Utah has constructed two arrays of low angular resolution telescopes in the desert, shown in Fig. 8.15. Cosmic-rays impacting the upper atmosphere produce brief pulses of light as they strike molecules in the air. The faint ultraviolet flashes are only an indirect measurement of particles with energies of up to 10^{20} eV, far in excess of energies which can be

---

[260] A description of the observatory is given in *Sky & Telescope*, October, 1967, p. 204. The instrument has also been called the Fly's Eye telescope.

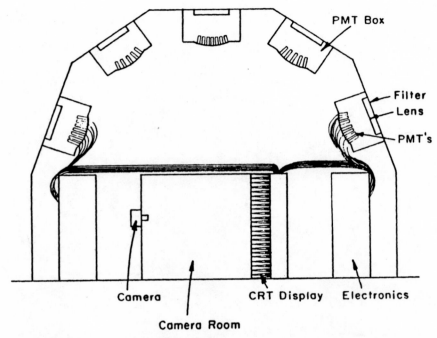

Figure 8.14. Cross-section of the Cornell Cosmic House. Drawing courtesy of Dr Peter Landecker.

generated in laboratories. Detection of cosmic-ray events is accomplished with 103 telescopes, each 1.5 m (59 in) in aperture. The telescopes are arranged in two arrays of 67 and 36 telescopes with about 3 km (1.9 mi) between the groups. Each telescope stares at one fixed position in the sky. At the focus of each telescope is a cluster of 12 to 14 photomultiplier tubes. Each tube looks at a patch of sky about five degrees in diameter. There are 880 phototubes in the larger array and 440 in the smaller one. Because of the method of detecting motion by watching the flash of light progress from one phototube to another, the array has been dubbed the "Fly's Eye".[261]

The telescopes are crude by most astronomical standards. Telescope tubes are made from culverts. The mirrors, fabricated by heat slumping 0.64 cm (0.25 in) thick plate glass over a curved mold, have only about one quarter degree of angular resolution and the spot size for stellar images is about 2.54 cm (1 in) in diameter.

As a cosmic-ray ionizes a trail of air particles several adjacent phototubes will detect the track and report the location of a cosmic-ray shower to a central computer.

[261] For a description of the array, see *Sky & Telescope*, September, 1981, p. 209 and for a discussion of cosmic rays measured by this and several other instruments see *Sky & Telescope*, August, 1985, p. 104. A similar array of photodetectors is being built by a consortium of researchers from Japan, Australia and New Zealand. This JANZOS array is described in *Sky & Telescope*, November, 1979, p. 455.

Figure 8.15. Array of telescopes used to detect cosmic-rays. Photo courtesy of University of Utah, Cosmic Ray Group.

The separation between the two arrays allows triangulation calculations to determine the height and trajectory of the particle. The system of telescopes can operate on any night when the Moon is not in the sky, weather permitting. While a study of charged particles does not reveal the astronomical source because the particle is bent in its trajectory by the galactic magnetic field, neutral particles will reveal their sources. Recently one of the sources of cosmic-rays on the order of 10^{18} electron volts has been identified as near Cygnus X-3.

Each of the separate optical systems in the gamma-ray and cosmic-ray detector telescopes ended in an individual photomultiplier. The data from each channel were not combined or correlated with other channels until they reached a computer which stored the information. Even at that point, only time and position data were correlated in order to find a faint streak of light crossing the sky. There are other, more subtle pieces of information in light reaching the telescopes but the data could be interpreted only by combining the light from two or more telescopes optically, not electronically. Since these "light-bucket" telescopes lack the high optical quality to perform interferometry, the technique is not viable for this experiment.

The technique of coherently combining the outputs of several individual telescopes via interferometry, originally pioneered in radio astronomy and now also applied to infrared and even some visible light telescopes, has become a powerful tool for astronomers. While it is an unusual technique of observation, the majority of telescopes employed in its use have been of classical design. A typical example is the National Radio Astronomy Observatory's Very Large Array (VLA), located near Socorro, New Mexico. The hardwired phased data processors, computers, control systems and kilometers of precision waveguides linking the elements advanced the state of the art in their design and conception. The problems of knowing where each antenna is located precisely with respect to the others is staggering and the thought of controlling 27 individual telescopes at once is enough to give any telescope operator pause. Alas, the entire wonderful system including acres of computers doesn't have a single unusual telescope. Each of the 25 m aperture radio telescopes is a classical Cassegrain reflector on an alt-az mount. The system probably doesn't even deserve mention in a book on unusual telescopes.[262]

The technique of combining signals interferometrically is much more difficult in the visible light region than in radio. This is because the precise locations of each of the optical elements must be controlled or known to within less than a wavelength of the signal received. In radio astronomy, the distances between telescopes in the array must be known to within a centimeter or so. This is a challenging but not impossible task when the telescopes are separated by a few kilometers. In the optical region, however, telescopes must be positioned to within a wavelength of light. If they cannot be positioned then optical path length compensation must be made where the light beams are combined and the compensation must be controlled to within a wavelength of light. The CERGA interferometer, shown in Fig. 4.13, uses a moving table to combine the light beams. A slightly longer wavelength experiment, the Infrared Spatial Interferometer shown in Fig. 5.7, uses similar techniques.

Small telescopes

Many astronomers feel their telescopes are too small. Indeed, "aperture envy" and "aperture fever" are common ailments among observers. There are, however, valid reasons for trying to make a telescope smaller. The most obvious is to make it more portable. If the telescope is to be transported in, for instance, a rocket then a few extra kilograms of mass or centimeters of size translates to thousands of dollars in fuel costs or, in the case of sub-orbital ballistic flights, precious seconds of observing time. Thousands of engineering manhours have gone into trimming and slimming optical systems for rockets.

Those of us who travel on airplanes to observe occultations and eclipses require small but powerful telescopes. My own favorite is a 90 mm aperture f/11 Maksutov telescope with an image-intensified CCD camera, video tape recorder and TV monitor. The whole package, including batteries and a small timing computer and

[262] A description of the VLA can be found in *Sky & Telescope*, June, 1975, p. 344, November, 1976, p. 320 and December, 1980, p. 473. An expansion into the Very Long Baseline Array (VLBA) is in the issue of June, 1985, p. 490.

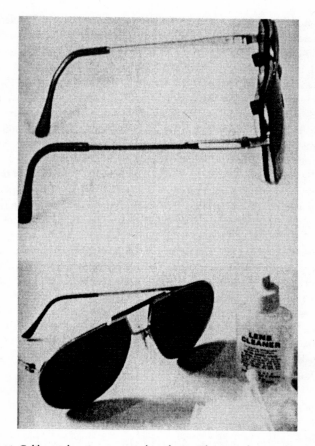

Figure 8.16. Galilean telescopes mounted in glasses. Photograph courtesy of Edwards Optical Company.

radio receiver, fit nicely under a standard airliner seat so that my precious observing equipment is never at the mercy of the baggage handlers. The trick here is in using an image intensifier instead of a larger telescope. The CCD camera would have the same sensitivity if I eliminated the intensifier and used a 35.15 cm (14 in) aperture telescope. International airfreight charges for that telescope and its mount run about twice the normal coach-class ticket price. It doesn't take many trips to realize that small is, indeed, beautiful.

Wesley N. Lindsay of San Jose, California, has designed some very small Schmidt-type telescopes for use as finders on larger instruments. The telescopes, ranging from 2.54 cm (1 in) in aperture, are related to the Mersenne telescope concept in that they do not use an eyepiece. The spherical primary and secondaries are ground so that the rays reflected from the secondary mirror form parallel bundles and can be focused by the eye without the aid of an eyepiece. Since there is no baffling in the telescope and the eye is placed just behind the primary, as in a classical Schmidt-Cassegrain, the field of view outside the magnified portion of the image is filled with stray light. This stray light, however, is not bothersome since it is just the unmagnified sky view of the area adjacent to the magnified field of view. While these rays pass through the Schmidt corrector plate, they are relatively undisturbed by the plate and form an unmagnified "finder" view around the magnified image. Thus, the system functions as a double field telescope. While there is some blending of the magnified and unmagnified views, a judicious placement of a black ring around the secondary will eliminate this.[263]

One of the smallest conventional telescopes currently manufactured is made for people with special vision problems. There is a class of low vision defects which can be aided by superimposing a magnified image on the normal scene. Miniature Galilean telescopes which magnify by about 2.5 × are mounted in conventional glasses frames as shown in Fig. 8.16. These telescopes are about 1 cm long and have a clear aperture of about 0.3 cm. The reason for using a Galilean optical design rather than a more conventional refractor is so that the image is erect. The viewer sees a magnified image near the center of his field of view but he also sees an unmagnified image at the periphery of his vision as he "looks around" the small telescopes. This produces an effect much like the double field telescopes discussed earlier. Interpreting the image takes a bit of getting used to but after a while, wearers of these glasses can drive a car, which they could not do without the glasses-mounted telescopes. They are made by the Edwards Optical Company of Virgina Beach, Virginia. Similar glasses have been tested by military pilots for target spotting.

Small aperture telescopes exist in many places in industry. They are more related to the periscope and microscope and are generally used to peer inside jet engines or down oil wells. While they are called telescopes, they don't have an astronomical application. There is, however, an application for very small apertures and that is to prevent too much starlight from entering the telescope. Too much light isn't usually a problem in astronomy but researchers studying the solar-stellar connection have to deal with it. In order to relate other stars to our own Sun they study both. Sensitive spectrographs have been built to study stars similar to the Sun. In order to make comparisons, solar spectra are also taken. The spectrographs, however, were built to record faint stars using 2 m class telescopes. If the telescopes were pointed to the Sun, they would probably melt the spectrographs. Thus, an alternative optical feed system is used. A fiber optic attached to an equatorial mount is pointed at the Sun and the output of the fiber optic is coupled to the spectrograph. The solar telescope shown in Fig. 8.17 was developed by the High Altitude Observatory under the sponsorship of the National Science Foundation. The telescope is mounted on the side of a dome at Lowell Observatory. Its 200 μm aperture is aimed at the Sun using a manual declination drive and an electric right ascension drive.

[263] Lindsay's patented double field Schmidt finder scopes are described at the end of an article by him appearing in *Sky & Telescope*, February, 1965, p. 112.

Figure 8.17. Fiber optic solar telescope used for comparison with stellar spectra. Photo courtesy of Richard Fisher, High Altitude Observatory.

9

Whimsy

As with any study, there are always a few items which don't even fit the "miscellaneous" category. They are often the product of a fertile sense of humor coupled with a cloudy, boring night. Such photographs may show completely normal astronomical activities which, when given appropriate captions, take on a whimsical meaning or they may include unconventional uses of astronomical equipment. It should be noted that the noble aspirations and high ideals of Astronomy do not make it immune from the lowest form of humor, the pun. The acknowledged grand master of truly vile astronomical puns is Dr Clyde Tombaugh and I've always wondered if some of his strange designs, such as the lawnmower-mounted telescope shown in Fig. 5.1, weren't executed with his tongue planted firmly in his cheek.

With its emphasis on optics, a significant percentage of astronomical humor is visual puns. Several years ago my wife made a hat for me with the letters "LENS" embroidered on the front. I often wear it to gatherings of telescope designers. It usually takes an observer a couple of minutes to realize that it's a "lens cap". One year at the Riverside Telescope Makers Conference I was giving a talk when, 20 minutes into the presentation, I heard a pained moan from the middle of the audience. Some people take a little longer to get it.[264]

Although the fork-mounted telescope was constructed as a joke based on a suggestion by David Levy, it does use a Galilean design with a magnification of $3 \times$. The telescope, shown in Fig. 9.1, sits on a shelf below the window in my den. Being an avid airplane watcher, it is often the only optical system handy when I want a closer look at some passing "bird". Thus, it has been used quite a bit more as a telescope than as a visual joke.

The telescope-driven clock is a modern model of an antique verge and foliot escapement clock, shown in Fig. 9.2. During a house move several years ago the driving weights were lost and, in order to make the instrument look more presentable, two small telescopes were used instead. This clock is so inaccurate that it has only an hour hand. It works better as a visual joke than as a timepiece.

The David Levy telescope collection is a unique assortment of astronomical

---

[264] As a punishment for inflicting this humor on the world, Richard Berry printed a picture of it in *Telescope Making*, No. 16, Summer, 1982, p. 30.

Figure 9.1. Fork-mounted telescope.

miscellanea. It includes serious antique Clark refractors, orphaned telescopes and a variety of lending telescopes, some made especially for use by children. There are also some oddball machines such as Mintaka, the talking telescope,[265] and a small telescope mounted on a model train.[266] One of my favorites is the beer can telescope, shown in Fig. 9.3. The primary mirror is made from a ladies' makeup compact mirror. The sighting device is the pull tab from the top. The telescope really does work, although it is not quite equal to multi-meter class instruments used on remote mountain tops. Perhaps this is because it is most often found in the parking lots of bars.

No, the people in Fig. 9.4 aren't performing some exotic dance. Using one hand to form a pinhole and the other to make a projection screen, an image of the partially eclipsed Sun was studied during the 1980 total solar eclipse in Kenya. The 0.25 cm (0.098 in) aperture f/200 optical system yields a reasonable image of the progress of the eclipse. This telescope is easy to use, inexpensive and portable. The tracking system, however, leaves something to be desired.

As an example of a normal astronomical activity given extraordinary meaning,

[265] The talking telescope is a 10.16 cm (4 in) refractor which has been folded with two flats and fitted into a speaker box with a tape recorder. It has been used educationally. See *Telescope Making*, No. 5, Fall, 1979, p. 8.

[266] The collection, including the model train telescope, is shown in *Sky & Telescope*, April, 1982, p. 401.

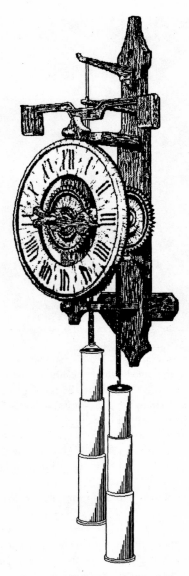

Figure 9.2. Telescope-driven clock.

Figure 9.3. Telescope made from a beer can by David Levy.

the photo in Fig. 9.5 was taken while moving the 63.5 cm (25 in) satellite-tracking telescope which is also shown in Fig. 1.15. Don Pedrick, the technician pictured, had climbed onto the truck to adjust the dust cover on the telescope when the picture was taken. The lettering on the telescope tube was added in the darkroom.

While attending the Riverside Telescope Makers Conference, I discovered that I'd forgotten my shaving mirror. Kevin Medlock of the San Jose Astronomical Association was kind enough to lend me a reasonable quarter-wave substitute, as shown in Fig. 9.6. Somebody suggested I avoid the use of messy shaving cream and a dangerous razor by waiting for the Sun to rise and standing at the focus to burn the whiskers off. Luckily, it was cloudy that morning.

Figure 9.4. I am a telescope. Photo by Bob Fingerhut of the San Jose Astronomical Association of Cheryl McConnell and Pete Manly.

Figure 9.5. The Great Pedrick. Photo by Jon Sellers.

Figure 9.6. A 46 cm (18 in) f/3 shaving mirror. Photo by Kevin Medlock.

Figure 9.7. Freemont Peak at Sunset. California amateur astronomers set up their telescopes in anticipation of clear, dark skies.

While there is a place for whimsy in astronomy, most of the telescopes shown in this book were created for serious purposes. Each of the instruments was designed by careful, thoughtful craftsmen who balanced the observing problem, the state of the art in materials and the available resources to come up with the very best telescope possible for their situation. More aperture, shiny knobs and electronics don't necessarily make one telescope better than another. One telescope may be bigger but a smaller one is more portable. Almost everybody has a telescope that looks better than my scratched, dented and highly modified occultation telescope. In the end, however, all telescopes look alike in the dark. What matters is the image you see through them.

Index

Author's note on the index; I appreciate a good index in a book but really good indices are hard to come by. The index is basically a word association game played between the author and the reader. If you are looking for a specific telescope or a class of telescopes and you don't find a reasonable entry in the index then please drop me a line at the address listed in the Preface. I would appreciate it and future readers would too.